Student Solutions Manual

for

Moore, McCabe, and Craig's
Introduction to the Practice of Statistics

Seventh Edition

Darryl K. Nester
Bluffton University

W. H. Freeman and Company
New York

©2012 by W.H. Freeman and Company

ISBN-10: 1-4292-7371-2
ISBN-13: 978-1-4292-7371-8

All rights reserved.

Printed in the United States of America

First printing

W.H. Freeman and Company
41 Madison Avenue
New York, NY 10010
Houndmills, Basingstoke RG21 6XS England

www.whfreeman.com

CONTENTS

About these solutions

The solutions that follow were prepared by Darryl Nester. In some cases, solutions were based on those prepared for earlier editions of *IPS*. Jackie Miller reviewed the solutions, especially focusing on those which were new (or revised) from previous editions. In spite of the care that went into that process, I might have missed a subtle change in an exercise that should have resulted in a change in the solution. Should you discover any errors or have any comments about these solutions (or the odd answers, in the back of the text), please report them to me:

> Darryl Nester
> Bluffton University
> Bluffton, Ohio 45817
> e-mail: nesterd@bluffton.edu
> WWW: www.bluffton.edu/~nesterd

The Student Solutions Guide is intended to provide complete solutions to all of the odd-numbered exercises, except for those for which no definitive answer can be given, such that those that ask students to look up and choose data on the Internet. In the case of simple exercises, this might mean that the "solution" is identical to the answer given in the back of the text. For more complicated exercises, the solution includes more detail about the *process* of determining the answer.

To create this guide, the odd-numbered solutions were extracted from the Instructor's Guide (which contains solutions to *all* exercises). **Some editing was done after this extraction, but a few comments aimed at instructors still remain in these solutions.**

These solutions were prepared using:

- For typesetting: TEX — specifically, Textures, from Blue Sky Software.

- For the graphs: DeltaGraph (SPSS), Adobe Illustrator, and R.

- For statistical analysis: primarily Minitab and Excel, along with SAS (for Chapter 14) and some freeware and shareware software (R, G•Power, and GLMStat).

The solutions given to the applet exercises, and the sample output screens, were based on the current versions of the applets at the time the solutions were written. As revisions are made to these applets, the appearance of the output screens (and in some cases, the answers) may change. Screenshots were taken on a computer running Macintosh OS X, but that should have no significant impact on the appearance.

Using the table of random digits

Grading SRSs chosen from Table B is complicated by the fact that students can find some creative ways to (mis)use the table. Some approaches are not mistakes, but may lead to different students having different "right" answers. Correct answers will vary based on

- The line in the table on which they begin (you may want to specify one if the text does not).

- Whether they start with, e.g., 00 or 01.

- Whether they assign multiple labels to each unit.

- Whether they assign labels across the rows or down the columns (nearly all lists in the text are alphabetized down the columns).

Some approaches can potentially lead to wrong answers. Mistakes to watch out for include the following:

- They may forget that all labels must be the same length (e.g., assigning labels such as $0, 1, 2, \ldots, 9, 10, \ldots$ rather than $00, 01, 02, \ldots$).

- In assigning multiple labels, they may not give the same number of labels to all units. For example, if there are 30 units, they may try to use up all the two-digit numbers, thus assigning four labels to the first 10 units and only three to the remaining 20.

As an alternative to using the random digits in Table B, students can pick a random sample by generating (pseudo)-random numbers, using software (like Excel) or a calculator. With many, if not all, calculators, the sequence of random numbers produced is determined by a "seed value" (which can be specified by the user). Rather than pointing students to a particular line of Table B, you could specify a seed value for generating random numbers, so that all students would obtain the same results (if all are using the same model of calculator).

On a TI-83, for example, after executing the command `0→rand`, the `rand` command will produce the sequence (rounded to four decimals) 0.9436, 0.9083, 0.1467, . . ., while `1→rand` initiates the sequence 0.7456, 0.8559, 0.2254, . . . So to choose, say, an SRS of size 10 from 30 subjects, use the command `0→rand` to set the seed, and then type `1+30*rand`, and press ENTER repeatedly. Ignoring the decimal portion of the resulting numbers, this produces the sample

$$29, \ 28, \ 5, \ 15, \ 13, \ 23, \ 2, \ 11, \ 30, \ 7$$

(Generally, to generate random numbers from 1 to n, use the command `1+n*rand` and ignore the decimal portion of the result.)

Using statistical software

The use of computer software or a calculator is a must for all but the most cursory treatment of the material in this text. Be aware of the following considerations:

- *Standard deviations:* Students may be confused by software that gives both the so-called "sample standard deviation" (the one used in the text) and the "population standard deviation" (dividing by n rather than $n - 1$). Symbolically, the former is usually given as s and the latter as σ (sigma), but the distinction is not always clear. For example, many computer spreadsheets have a command such as "STDEV(...)" to compute standard deviations, but you may need to check the manual to find out which kind it is.

    As a quick check: For the numbers 1, 2, 3, $s = 1$ while $\sigma \doteq 0.8165$. In general, if two values are given, the larger one is s and the smaller is σ. If only one value is given, and it is the wrong one, use the relationship $s = \sigma \sqrt{\dfrac{n}{n - 1}}$.

- *Stemplots:* The various choices one can make in creating a stemplot (e.g., rounding or truncating the data) have already been mentioned. Minitab opts for truncation over rounding, so all of the solutions in this guide show truncated-data stemplots (except for exercises that instructed students to round). This usually makes little difference in the overall appearance of the stemplot.

- *Significant digits in these solutions:* Most numerical answers in these solutions (and in the odd-numbered answers in the back of the text) are reported to four significant figures. In many cases, that is an absurd overrepresentation of the accuracy of those numbers, but those digits are provided to give students a better "check" on their answers.

  This extra accuracy is a double-edged sword, however, as a student might have a correct answer that does not agree with all the digits in the printed answer. This might occur (rarely, I hope) because my answer is wrong, but it may be due to rounding, differences in software accuracy, or use of an approximation. For example, in reporting binomial probabilities for exercises in Chapter 5, I have listed four or five answers for some problems: exact answers and Normal approximations (with or without the continuity correction, and computed with software or using Table A). In the back-of-the-text answers, only one answer is given, so that students may have to be satisfied with being "close."

- *Quartiles and five-number summaries:* Methods of computing quartiles vary between different packages. Some use the approach given in the text (that is, Q_1 is the median of all the numbers below the location of the overall median, etc.), while others use a more complicated approach. For the numbers 1, 2, 3, 4, for example, we would have $Q_1 = 1.5$ and $Q_3 = 3.5$, but Minitab reports these as 1.25 and 3.75, respectively, while Excel reports 1.75 and 3.25.

  In these solutions (and the odd-numbered answers in the back of the text), I opted to report five-number summaries are they would be found using the text's method. Because I used Minitab for most of the analysis in these solutions, I wrote a Minitab macro to compute quartiles the *IPS* way. This and other macros are available on my Web site.

- *Boxplots:* Some programs that draw boxplots use the convention that the "whiskers" extend to the lower and upper deciles (the 10th and 90th percentiles) rather than to the minimum and maximum. The information conveyed by such a boxplot is essentially the same as by those described and used in the text; if students use such software, point out that they can still draw the same conclusions about the underlying distribution.

1.25. **(a)** & **(b)** Both bar graphs are shown below. **(c)** The ordered bars in the graph from (b) make it easier to identify those materials that are frequently recycled and those that are not. **(d)** Each percent represents part of a different whole. (For example, 2.6% of *food scraps* are recycled; 23.7% of *glass* is recycled, etc.)

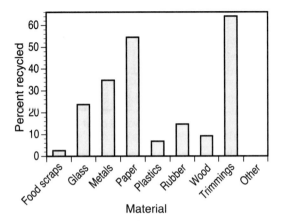 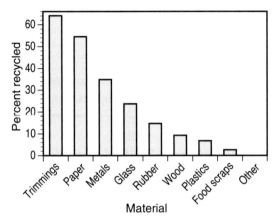

1.27. The two bar graphs are shown below.

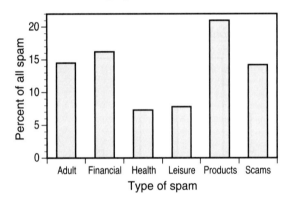 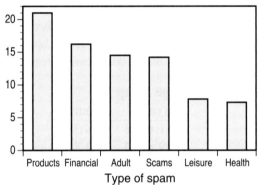

1.29. **(a)** Most countries had moderate (single- or double-digit) increases in Facebook usages. Chile (2197%) is an extreme outlier, as are (maybe) Venezuela (683%) and Colombia (246%). **(b)** In the stemplot on the right, Chile and Venezuela have been omitted, and stems are split five ways. **(c)** One observation is that, even without the outliers, the distribution is right-skewed. **(d)** The stemplot can show some of the detail of the low part of the distribution, if the outliers are omitted.

```
0 | 000
0 | 2333
0 | 4444
0 | 6
0 | 99
1 |
1 | 33
1 |
1 |
1 |
2 |
2 |
2 | 4
```

1.31. (a) The luxury car bar graph is below on the left; bars are in decreasing order of size (the order given in the table). **(b)** The intermediate car bar graph is below on the right. For this stand-alone graph, it seemed appropriate to re-order the bars by decreasing size. Students may leave the bars in the order given in the table; this (admittedly) might make comparison of the two graphs simpler. **(c)** The graph on the right is one possible choice for comparing the two types of cars: for each color, we have one bar for each car type.

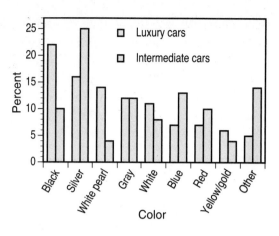

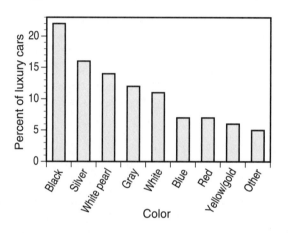

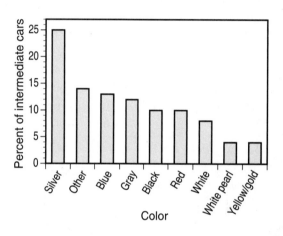

1.33. Shown is the stemplot; as the text suggests, we have trimmed numbers (dropped the last digit) and split stems. 359 mg/dl appears to be an outlier. Overall, glucose levels are not under control: Only 4 of the 18 had levels in the desired range.

```
0 | 799
1 | 0134444
1 | 5577
2 | 0
2 | 57
3 |
3 | 5
```

1.35. The distribution is roughly symmetric, centered near 7 (or "between 6 and 7"), and spread from 2 to 13.

1.37. To display the distribution, use either a stemplot or a histogram. DT scores are skewed to the right, centered near 5 or 6, spread from 0 to 18. There are no outliers. We

```
0 | 00000000000000000000000000000000001111111111111111111
0 | 2222222222222222233333333333333333333333
0 | 444444444444444444445555555555555555555
0 | 66666666666666666666777777777777
0 | 88888888888888899999999999999999
1 | 00000000000111111111
1 | 2222222222223333333333333
1 | 444444455
1 | 66666777
1 | 8
```

might also note that only 11 of these 264 women (about 4%) scored 15 or higher.

1.39. Graph (a) is studying time (Question 4); it is reasonable to expect this to be right-skewed (many students study little or not at all; a few study longer).

Graph (d) is the histogram of student heights (Question 3): One would expect a fair amount of variation but no particular skewness to such a distribution.

The other two graphs are (b) handedness and (c) gender—unless this was a particularly unusual class! We would expect that right-handed students should outnumber lefties substantially. (Roughly 10 to 15% of the population as a whole is left-handed.)

1.41. (a) Not only are most responses multiples of 10; many are multiples of 30 and 60. Most people will "round" their answers when asked to give an estimate like this; in fact, the most striking answers are ones such as 115, 170, or 230. The students who claimed 360 minutes (6 hours) and 300 minutes (5 hours) may have been exaggerating. (Some students might also "consider suspicious" the student who claimed to study 0 minutes per night. As a teacher, I can easily believe that such students exist, and I suspect that some of your students might easily accept that claim as well.) **(b)** The stemplots suggest that women (claim to) study more than men. The approximate centers are 175 minutes for women and 120 minutes for men.

Women		Men
	0	033334
96	0	66679999
22222221	1	2222222
888888888875555	1	558
4440	2	00344
	2	
	3	0
6	3	

1.43. (a) There are four variables: GPA, IQ, and self-concept are quantitative, while gender is categorical. (OBS is not a variable, since it is not really a "characteristic" of a student.) **(b)** Below. **(c)** The distribution is skewed to the left, with center (median) around 7.8. GPAs are spread from 0.5 to 10.8, with only 15 below 6. **(d)** There is more variability among the boys; in fact, there seems to be a subset of boys with GPAs from 0.5 to 4.9. Ignoring that group, the two distributions have similar shapes.

0	5
1	8
2	4
3	4689
4	0679
5	1259
6	0112249
7	22333556666666788899
8	0000222223347899
9	002223344556668
10	01678

Female		Male
	0	5
	1	8
	2	4
4	3	689
7	4	069
952	5	1
4210	6	129
98866533	7	223566666789
997320	8	0002222348
65300	9	2223445668
710	10	68

1.45. Stemplot at right, with split stems. The distribution is skewed to the left, with center around 59.5. Most self-concept scores are between 35 and 73, with a few below that, and one high score of 80 (but not really high enough to be an outlier).

2	01
2	8
3	0
3	5679
4	02344
4	6799
5	1111223344444
5	556668899
6	00001233344444
6	55666677777899
7	0000111223
7	
8	0

1.47. The total for the 24 countries was 897 days, so with Suriname, it is $897 + 694 = 1591$ days, and the mean is $\bar{x} = \frac{1591}{25} = 63.64$ days.

1.49. To find the ordered list of times, start with the 24 times in Example 1.23, and add 694 to the end of the list. The ordered times (with median highlighted) are

$$4, \ 11, \ 14, \ 23, \ 23, \ 23, \ 23, \ 24, \ 27, \ 29, \ 31, \ 33, \ \boxed{40},$$
$$42, \ 44, \ 44, \ 44, \ 46, \ 47, \ 60, \ 61, \ 62, \ 65, \ 77, \ 694$$

The outlier increases the median from 36.5 to 40 days, but the change is much less than the outlier's effect on the mean.

1.51. In order, the scores are:

$$55, \ 73, \ 75, \ 80, \ \boxed{80}, \ \boxed{85}, 90, \ 92, \ 93, \ 98$$

The middle two scores are 80 and 85, so the median is $M = \dfrac{80 + 85}{2} = 82.5$.

1.53. The maximum and minimum can be found by inspecting the list. The sorted list (with quartile and median locations highlighted) is

1	2	2	3	4	9	9	9	11	19
19	25	30	35	40	44	48	51	52	54
55	56	57	59	64	67	68	73	73	75
75	76	76	77	80	88	89	90	102	103
104	106	115	116	118	121	126	128	137	138
140	141	143	148	148	157	178	179	182	199
201	203	211	225	274	277	289	290	325	367
372	386	438	465	479	700	700	951	1148	2631

This confirms the five-number summary (1, 54.5, 103.5, 200, and 2631 seconds) given in Example 1.26. The sum of the 80 numbers is 15,726 seconds, so the mean is $\bar{x} = \frac{15,726}{80} = 196.575$ seconds (the value 197 in the text was rounded).

    Note: *The most tedious part of this process is sorting the numbers and adding them all up. Unless you really want to confirm that your students can sort a list of 80 numbers, consider giving the students the sorted list of times, and checking their ability to identify the locations of the quartiles.*

1.55. Use the five-number summary from the solution to Exercise 1.54:

$$\text{Min} = 55, \ \ Q_1 = 75, \ \ M = 82.5, \ \ Q_3 = 92, \ \ \text{Max} = 98$$

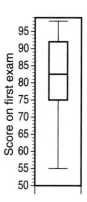

1.57. The variance *can* be computed from the formula $s^2 = \dfrac{1}{n-1}\sum(x_i - \bar{x})^2$; for example, the first term in the sum would be $(80 - 82.1)^2 = 4.41$. However, in practice, software or a calculator is the preferred approach; this yields $s^2 = \dfrac{1416.9}{9} = 157.4\overline{3}$ and $s = \sqrt{s^2} \doteq 12.5472$.

1.59. Without Suriname, the quartiles are 23 and 46.5 days; with Suriname included, they are 23 and 53.5 days. Therefore, the *IQR* increases from 23.5 to 30.5 days—a much less drastic change than the change in s (18.6 to 132.6 days).

1.61. (a) Use a stemplot or histogram. **(b)** Because the distribution is skewed, the five-number summary is the best choice; in millions of dollars, it is

Min	Q_1	M	Q_3	Max
3338	4589	7558.5	13,416	66,667

Some students might choose the less-appropriate summary: $\bar{x} \doteq 12,144$ and $s \doteq 12,421$ million dollars. **(c)** For example, the distribution is sharply right-skewed. (This is not surprising given that we are looking at the top 100 companies; the top fraction of most distributions will tend to be skewed to the right.)

```
0 | 3333333333333333333344444444444444
0 | 55555555566666667777777777778888889
1 | 00001112223333333
1 | 79
2 | 01111233
2 | 559
3 | 114
3 | 5
4 |
4 |
5 | 3
5 | 99
6 |
6 | 6
```

1.63. (a) $\bar{x}$ changes from 4.76% (with) to 4.81% (without); the median (4.7%) does not change. **(b)** s changes from 0.7523% to 0.5864%; Q_1 changes from 4.3% to 4.35%, while $Q_3 = 5\%$ does not change. **(c)** A low outlier decreases $\bar{x}$; any kind of outlier increases s. Outliers have little or no effect on the median and quartiles.

1.65. Use a small data set with an odd number of points, so that the median is the middle number. After deleting the lowest observation, the median will be the average of that middle number and the next number after it; if that latter number is much larger, the median will change substantially. For example, start with 0, 1, **2**, 998, 1000; after removing 0, the median changes from 2 to 500.

1.67. (a) The distribution is left-skewed. While the skew makes the five-number summary is preferable, some students might give the mean/standard deviation. In ounces, these statistics are:

$\bar{x}$	s	Min	Q_1	M	Q_3	Max
6.456	1.425	3.7	4.95	6.7	7.85	8.2

(b) The numerical summary does not reveal the two weight clusters (visible in a stemplot or histogram). **(c)** For small potatoes (less than 6 oz), $n = 8$, $\bar{x} = 4.662$ oz, and $s = 0.501$ oz. For large potatoes, $n = 17$, $\bar{x} = 7.300$ oz, and $s = 0.755$ oz. Because there are clearly two groups, it seems appropriate to treat them separately.

```
3 | 7
4 | 3
4 | 7777
5 | 23
5 |
6 | 0033
6 | 7
7 | 03
7 | 668899999
8 | 2
```

1.69. (a) The five-number summary is Min $= 0$ mg/l, $Q_1 = 0$ mg/l, $M = 5.085$ mg/l, $Q_3 = 9.47$ mg/l, Max $= 73.2$ mg/l. **(b) & (c)** The boxplot and histogram are shown below. (Students might choose different interval widths for the histogram.) **(d)** Preferences will vary. Both plots reveal the sharp right-skew of this distribution, but because Min $= Q_1$, the boxplot looks somewhat strange. The histogram seems to convey the distribution better.

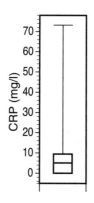
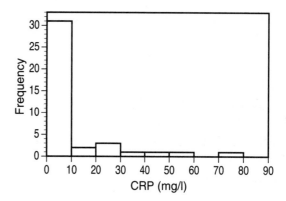

1.71. (a) The five-number summary (in units of μmol/l) is Min $= 0.24$, $Q_1 = 0.355$, $M = 0.76$, $Q_3 = 1.03$, Max $= 1.9$. **(b) & (c)** The boxplot and histogram are shown below. (Students might choose different interval widths for the histogram.) **(d)** The distribution is right-skewed. A histogram (or stemplot) is preferable because it reveals an important feature not evident from a boxplot: This distribution has two peaks.

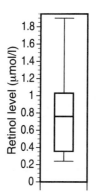
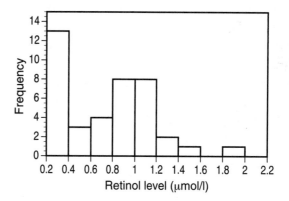

1.73. The distribution of household net worth would almost surely be strongly skewed to the right: Most families would generally have accumulated little or modest wealth, but a few would have become rich. This strong skew pulls the mean to be higher than the median.

1.75. The total salary is $690,000, so the mean is $\bar{x} = \frac{\$690,000}{9} \doteq \$76,667$. Six of the nine employees earn less than the mean. The median is $M = \$35,000$.

1.77. The total salary is now $825,000, so the new mean is $\bar{x} = \frac{\$825,000}{9} \doteq \$91,667$. The median is unchanged.

1.79. The quote describes a distribution with a strong right skew: Lots of years with no losses to hurricane ($0), but very high numbers when they do occur. For example, if there is one hurricane in a 10-year period causing $1 million in damages, the "average annual loss" for that period would be $100,000, but that does not adequately represent the cost for the year of the hurricane. Means are not the appropriate measure of center for skewed distributions.

1.81. (a) & (b) See the table on the right. In both cases, the mean and median are quite similar.

	$\bar{x}$	s	M
pH	5.4256	0.5379	5.44
Density	5.4479	0.2209	5.46

1.83. With only two observations, the mean and median are always equal because the median is halfway between the middle two (in this case, the only two) numbers.

1.85. (a) There are several different answers, depending on the configuration of the first five points. *Most students* will likely assume that the first five points should be distinct (no repeats), in which case the sixth point *must* be placed at the median. This is because the median of 5 (sorted) points is the third, while the median of 6 points is the average of the third and fourth. If these are to be the same, the third and fourth points of the set of six must both equal the third point of the set of five.

   The diagram below illustrates all of the possibilities; in each case, the arrow shows the location of the median of the initial five points, and the shaded region (or dot) on the line indicates where the sixth point can be placed without changing the median. Notice that there are four cases where the median does not change, regardless of the location of the sixth point. (The points need not be equally spaced; these diagrams were drawn that way for convenience.)

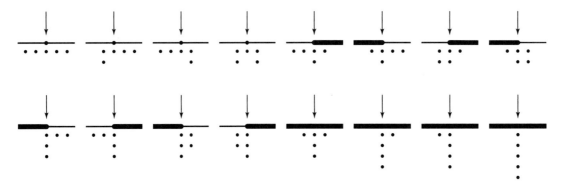

(b) Regardless of the configuration of the first five points, if the sixth point is added so as to leave the median unchanged, then in that (sorted) set of six, the third and fourth points must be equal. One of these two points will be the middle (fourth) point of the (sorted) set of seven, no matter where the seventh point is placed.

   Note: *If you have a student who illustrates all possible cases above, then it is likely that the student either (1) obtained a copy of this solutions manual, (2) should consider a career in writing solutions manuals, (3) has too much time on his or her hands, or (4) both 2 and 3 (and perhaps 1) are true.*

1.87. (a) The means and standard deviations (all in millimeters) are:

Variety	$\bar{x}$	s
bihai	47.5975	1.2129
red	39.7113	1.7988
yellow	36.1800	0.9753

```
bihai              red            yellow
46 | 3466789      37 | 4789      34 | 56
47 | 114          38 | 0012278   35 | 146
48 | 0133         39 | 167       36 | 0015678
49 |              40 | 56        37 | 01
50 | 12           41 | 4699      38 | 1
                  42 | 01
                  43 | 0
```

(b) *Bihai* and red appear to be right-skewed (although it is difficult to tell with such small samples). Skewness would make these distributions unsuitable for $\bar{x}$ and s.

1.89. The minimum and maximum are easily determined to be 1 and 12 letters, and the quartiles and median can be found by adding up the bar heights. For example, the first two bars have total height 22.3% (less than 25%), and adding the third bar brings the total to 45%, so Q_1 must equal 3 letters. Continuing this way, we find that the five-number summary, in units of letters, is:

$$\text{Min} = 1, \quad Q_1 = 3, \quad M = 4, \quad Q_3 = 5, \quad \text{Max} = 12$$

Note that even without the frequency table given in the data file, we could draw the same conclusion by estimating the heights of the bars in the histogram.

1.91. The simplest approach is to take (at least) six numbers—say, a, b, c, d, e, f in increasing order. For this set, $Q_3 = e$; we can cause the mean to be larger than e by simply choosing f to be *much* larger than e. For example, if all numbers are nonnegative, $f > 5e$ would accomplish the goal because then

$$\bar{x} = \frac{a+b+c+d+e+f}{6} > \frac{e+f}{6} > \frac{e+5e}{6} = e.$$

1.93. (a) One possible answer is 1, 1, 1, 1. **(b)** 0, 0, 20, 20. **(c)** For (a), any set of four identical numbers will have $s = 0$. For (b), the answer is unique; here is a rough description of why. We want to maximize the "spread-out"-ness of the numbers (which is what standard deviation measures), so 0 and 20 seem to be reasonable choices based on that idea. We also want to make each individual squared deviation—$(x_1 - \bar{x})^2$, $(x_2 - \bar{x})^2$, $(x_3 - \bar{x})^2$, and $(x_4 - \bar{x})^2$—as large as possible. If we choose 0, 20, 20, 20—or 20, 0, 0, 0—we make the first squared deviation 15^2, but the other three are only 5^2. Our best choice is two at each extreme, which makes all four squared deviations equal to 10^2.

1.95. The table on the right reproduces the means and standard deviations from the solution to Exercise 1.87 and shows those values expressed in inches. For each conversion, multiply by $39.37/1000 = 0.03937$ (or divide by 25.4—an inch is defined as 25.4 millimeters). For example, for the *bihai* variety, $\bar{x} = (47.5975 \text{ mm})(0.03937 \text{ in/mm}) = (47.5975 \text{ mm}) \div (25.4 \text{ mm/in}) = 1.874$ in.

	(in mm)		(in inches)	
Variety	$\bar{x}$	s	$\bar{x}$	s
bihai	47.5975	1.2129	1.874	0.04775
red	39.7113	1.7988	1.563	0.07082
yellow	36.1800	0.9753	1.424	0.03840

1.97. Convert from kilograms to pounds by multiplying by 2.2: $\bar{x} = (2.42 \text{ kg})(2.2 \text{ lb/kg}) \doteq 5.32$ lb and $s = (1.18 \text{ kg})(2.2 \text{ lb/kg}) \doteq 2.60$ lb.

1.99. There are 80 service times, so to find the 10% trimmed mean, remove the highest and lowest eight values (leaving 64). Remove the highest and lowest 16 values (leaving 48) for the 20% trimmed mean.

     The mean and median for the full data set are $\bar{x} = 196.575$ and $M = 103.5$ minutes. The 10% trimmed mean is $\bar{x}^* \doteq 127.734$, and the 20% trimmed mean is $\bar{x}^{**} \doteq 111.917$ minutes. Because the distribution is right-skewed, removing the extremes lowers the mean.

1.101. Take the mean plus or minus two standard deviations: $572 \pm 2(51) = 470$ to 674.

1.103. The z score is $z = \frac{620 - 572}{51} \doteq 0.94$.

1.105. Using Table A, the proportion below 620 ($z \doteq 0.94$) is 0.8264 and the proportion at or above is 0.1736; these two proportions add to 1. The graph on the right illustrates this with a single curve; it conveys essentially the same idea as the "graphical subtraction" picture shown in Example 1.36.

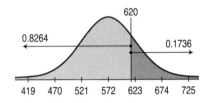

1.107. Using Table A, this ISTEP score should correspond to a standard score of $z \doteq 0.67$ (software gives 0.6745), so the ISTEP score (unstandardized) is $572 + 0.67(51) \doteq 606.2$ (software: 606.4).

1.109. Of course, student sketches will not be as neat as the curves on the right, but they should have roughly the correct shape. **(a)** It is easiest to draw the curve first, and then mark the scale on the axis. **(b)** Draw a copy of the first curve, with the peak over 20. **(c)** The curve has the same shape, but is translated left or right.

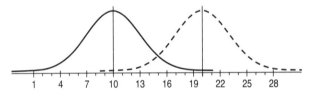

1.111. Sketches will vary.

1.113. (a) Ranges are given in the table on the right. In both cases, some of the lower limits are negative, which does not make sense; this happens because the women's distribution is skewed, and the men's distribution has an outlier. Contrary to the conventional wisdom, the men's mean is slightly higher, although the outlier is at least partly responsible for that. **(b)** The means suggest that Mexican men and women tend to speak more than people of the same gender from the United States.

	Women	Men
68%	8489 to 20,919	7158 to 22,886
95%	2274 to 27,134	−706 to 30,750
99.7%	−3941 to 33,349	−8,570 to 38,614

1.115. (a) We need the 5th, 15th, 55th, and 85th percentiles for a $N(0, 1)$ distribution. These are given in the table on the right. **(b)** To convert to actual scores, take the standard-score cut-off z and compute $10z + 70$. **(c)** Opinions will vary.

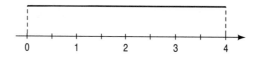

	Table A		Software	
	Standard	Actual	Standard	Actual
F	−1.64	53.6	−1.6449	53.55
D	−1.04	59.6	−1.0364	59.64
C	0.13	71.3	0.1257	71.26
B	1.04	80.4	1.0364	80.36

Note: *The cut-off for an A given in the previous solution is the* lowest *score that gets an A—that is, the point where one's grade drops from an A to a B. These cut-offs are the points where one's grade jumps* up. *In practice, this is only an issue for a score that falls exactly on the border between two grades.*

1.117. (a) The height should be $\frac{1}{4}$ since the area under the curve must be 1. The density curve is on the right. **(b)** $P(X \leq 1) = \frac{1}{4} = 0.25$. **(c)** $P(0.5 < X < 2.5) = 0.5$.

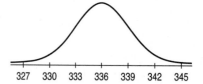

1.119. (a) Mean is C, median is B (the right skew pulls the mean to the right). **(b)** Mean A, median A. **(c)** Mean A, median B (the left skew pulls the mean to the left).

1.121. (a) The applet shows an area of 0.6826 between −1.000 and 1.000, while the 68–95–99.7 rule rounds this to 0.68. **(b)** Between −2.000 and 2.000, the applet reports 0.9544 (compared to the rounded 0.95 from the 68–95–99.7 rule). Between −3.000 and 3.000, the applet reports 0.9974 (compared to the rounded 0.997).

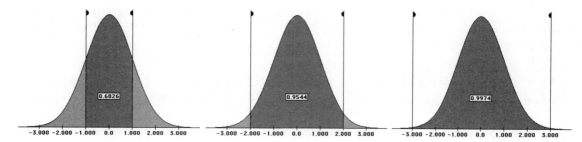

1.123. (a) 99.7% of horse pregnancies fall within three standard deviations of the mean: $336 \pm 3(3)$, or 327 to 325 days. **(b)** About 16% are longer than 339 days since 339 days or more corresponds to at least one standard deviation above the mean.

Note: *This exercise did not ask for a sketch of the Normal curve, but students should be encouraged to make such sketches anyway.*

1.125. The mean and standard deviation are $\bar{x} = 5.4256$ and $s = 0.5379$. About 67.62% ($71/105 \doteq 0.6476$) of the pH measurements are in the range $\bar{x} \pm s = 4.89$ to 5.96. About 95.24% (100/105) are in the range $\bar{x} \pm 2s = 4.35$ to 6.50. All (100%) are in the range $\bar{x} \pm 3s = 3.81$ to 7.04.

1.127. Using values from Table A:
 (a) $Z \leq -1.8$: 0.0359. (b) $Z \geq -1.8$:
0.9641. (c) $Z > 1.6$: 0.0548. (d) $-1.8 <$
$Z < 1.6$: $0.9452 - 0.0359 = 0.9093$.

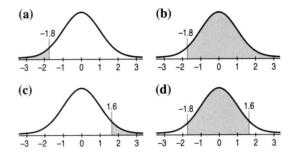

1.129. (a) $z = 0.3853$ has cumulative proportion 0.65 (that is, 0.3853 is the 65th percentile of the standard Normal distribution). (b) If $z = 0.1257$, then $Z > z$ has proportion 0.45 (0.1257 is the 55th percentile).

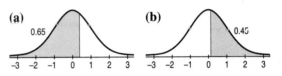

1.131. 130 is two standard deviations above the mean (that is, it has standard score $z = 2$), so about 2.5% of adults would score at least 130.

1.133. Jacob's score standardizes to $z = \frac{16 - 21.5}{5.4} \doteq -1.0185$, while Emily's score corresponds to $z = \frac{1020 - 1509}{321} \doteq -1.5234$. Jacob's score is higher.

1.135. Maria's score standardizes to $z = \frac{30 - 21.5}{5.4} \doteq 1.5741$, so an equivalent SAT score is $1509 + 1.5741 \times 321 \doteq 2014$.

1.137. Jacob's score standardizes to $z = \frac{19 - 21.5}{5.4} \doteq -0.4630$, for which Table A gives 0.3228. His score is the 32.3 percentile.

1.139. 1239 and below: The bottom 20% corresponds to a standard score of $z = -0.8416$, which in turn corresponds to a score of $1509 - 0.8416 \times 321 \doteq 1239$ on the SAT.

1.141. The quintiles of the SAT score distribution are $1509 - 0.8416 \times 321 = 1239$, $1509 - 0.2533 \times 321 = 1428$, $1509 + 0.2533 \times 321 = 1590$, and $1509 + 0.8416 \times 321 = 1779$.

1.143. For a Normal distribution with mean 46 mg/dl and standard deviation 13.6 mg/dl:
 (a) 40 mg/dl standardizes to $z = \frac{40 - 46}{13.6} \doteq -0.4412$. Using Table A, 33% of men fall below this level (software: 32.95%). (b) 60 mg/dl standardizes to $z = \frac{60 - 46}{13.6} \doteq 1.0294$. Using Table A, 15.15(c) Subtract the answers from (a) and (b) from 100%: Table A gives 51.85% (software: 51.88%), so about 52% of men fall in the intermediate range.

1.145. (a) About 5.2%: $x < 240$ corresponds to $z < -1.625$. Table A gives 5.16% for -1.63 and 5.26% for -1.62. Software (or averaging the two table values) gives 5.21%. (b) About 54.7%: $240 < x < 270$ corresponds to $-1.625 < z < 0.25$. The area to the left of 0.25 is 0.5987; subtracting the answer from part (a) leaves about 54.7%. (c) About 279 days or longer: Searching Table A for 0.80 leads to $z > 0.84$, which corresponds to $x > 266 + 0.84(16) = 279.44$. (Using the software value $z > 0.8416$ gives $x > 279.47$.)

1.147. **(a)** As the quartiles for a standard Normal distribution are ± 0.6745, we have $IQR = 1.3490$. **(b)** $c = 1.3490$: For a $N(\mu, \sigma)$ distribution, the quartiles are $Q_1 = \mu - 0.6745\sigma$ and $Q_3 = \mu + 0.6745\sigma$.

1.149. **(a)** The first and last deciles for a standard Normal distribution are ± 1.2816. **(b)** For a $N(9.12, 0.15)$ distribution, the first and last deciles are $\mu - 1.2816\sigma \doteq 8.93$ and $\mu + 1.2816\sigma \doteq 9.31$ ounces.

1.151. **(a)** The plot is reasonably linear except for the point in the upper right, so this distribution is roughly Normal, but with a high outlier. **(b)** The plot is fairly linear, so the distribution is roughly Normal. **(c)** The plot curves up to the right—that is, the large values of this distribution are larger than they would be in a Normal distribution—so the distribution is skewed to the right.

1.153. **(a)** All three quantile plots are below; the yellow variety is the nearest to a straight line. **(b)** The other two distributions are slightly right-skewed (the lower-left portion of the graph is somewhat flat); additionally, the *bihai* variety appears to have a couple of high outliers.

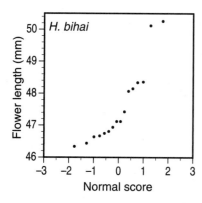
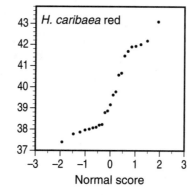
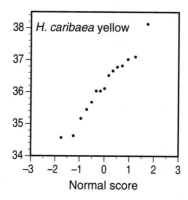

1.155. Shown are a histogram and quantile plot for one sample of 200 simulated uniform data points. Histograms will vary slightly but should suggest the density curve of Figure 1.34 (but with more variation than students might expect). The Normal quantile plot shows that, compared to a Normal distribution, the uniform distribution does not extend as low or as high (not surprising, since all observations are between 0 and 1).

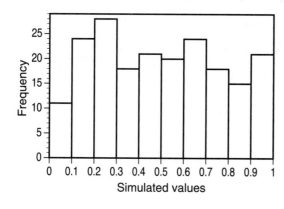
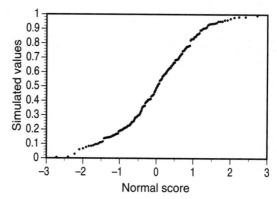

1.157. **(a)** The distribution appears to be roughly Normal. **(b)** One could justify using either the mean and standard deviation or the five-number summary:

$\bar{x}$	s	Min	Q_1	M	Q_3	Max
15.27%	3.118%	8.2%	13%	15.5%	17.6%	22.8%

(c) For example, binge drinking rates are typically 10% to 20%. Which states are high, and which are low? One might also note the geographical distribution of states with high binge-drinking rates: The top six states (Wisconsin, North Dakota, Iowa, Minnesota, Illinois, and Nebraska) are all adjacent to one another.

```
 8 | 28
 9 |
10 | 58
11 | 34
12 | 023689
13 | 015788
14 | 0077
15 | 13466889
16 | 01567
17 | 45677789
18 | 8
19 | 148
20 | 2
21 | 6
22 | 8
```

1.159. Students might compare color preferences using a stacked bar graph like that shown on the right, or side-by-side bars like those below. (They could also make six pie charts, but comparing slices across pies is difficult.) Possible observations: white is considerably less popular in Europe, and gray is less common in China.

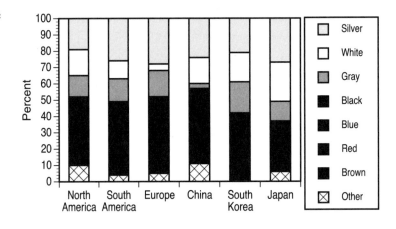

Note: *The orders of countries and colors is as given in the text, which is more-or-less arbitrary. (Colors are ordered by decreasing popularity in North America.)*

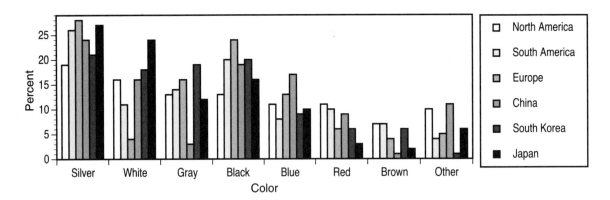

1.163. The distribution is somewhat right-skewed (although considerably less than the distribution with all countries) with only one country (Bosnia and Herzegovina) in the 20's. Because of the irregular shape, students might choose either the mean/standard deviation or the five-number summary:

$\bar{x}$	s	Min	Q_1	M	Q_3	Max
39.85	22.05	1.32	18.68	43.185	54.94	85.65

```
0 | 145789
1 | 23488889
2 | 5
3 | 0134467
4 | 124666669
5 | 022345688
6 | 223
7 | 026
8 | 15
```

1.165. The given description is true on the average, but the curves (and a few calculations) give a more complete picture. For example, a score of about 675 is about the 97.5th percentile for both genders, so the top boys and girls have very similar scores.

1.167. Shown is a stemplot; a histogram should look similar to this. This distribution is relatively symmetric apart from one high outlier. Because of the outlier, the five-number summary (in hours) is preferred:

   22 23.735 24.31 24.845 28.55

Alternatively, the mean and standard deviation are $\bar{x} = 24.339$ and $s = 0.9239$ hours.

```
22 | 013
22 | 7899
23 | 0000112222233344444
23 | 5556666666677777788888888999
24 | 00000011111111222222222233333333333444444
24 | 5555555666666666777777888888999999
25 | 00001111233344
25 | 56666889
26 | 2
26 | .56
27 | 2
27 |
28 |
28 | 5
```

1.169. Either a bar graph or a pie chart could be used. The given numbers sum to 66.7, so the "Other" category presumably includes the remaining 29.3 million subscribers.

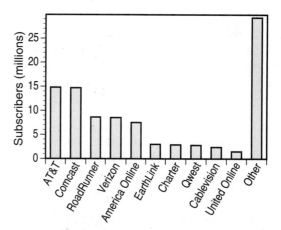

1.171. **(a)** For car makes (a categorical variable), use either a bar graph or pie chart. For car age (a quantitative variable), use a histogram, stemplot, or boxplot. **(b)** Study time is quantitative, so use a histogram, stemplot, or boxplot. To show change over time, use a time plot (average hours studied against time). **(c)** Use a bar graph or pie chart to show radio station preferences. **(d)** Use a Normal quantile plot to see whether the measurements follow a Normal distribution.

1.173. No, and no: It is easy to imagine examples of many different data sets with mean 0 and standard deviation 1—for example, $\{-1,0,1\}$ and $\{-2,0,0,0,0,0,0,0,2\}$.

Likewise, for any given five numbers $a \le b \le c \le d \le e$ (not all the same), we can create many data sets with that five-number summary, simply by taking those five numbers and adding some additional numbers in between them, for example (in increasing order): 10, __, 20, __, __, 30, __, __, 40, __, 50. As long as the number in the first blank is between 10 and 20, and so on, the five-number summary will be 10, 20, 30, 40, 50.

1.175. Bonds's mean changes from 36.56 to 34.41 home runs (a drop of 2.15), while his median changes from 35.5 to 34 home runs (a drop of 1.5). This illustrates that outliers affect the mean more than the median.

1	69
2	4
2	55
3	3344
3	77
4	02
4	5669
5	
5	
6	
6	
7	3

1.177. Results will vary. One set of 20 samples gave the results at the right (Normal quantile plots are not shown).

Theoretically, $\bar{x}$ will have a Normal distribution with mean 25 and standard deviation $8/\sqrt{30} \doteq 1.46$, so that about 99.7% of the time, one should find $\bar{x}$ between 20.6 and 29.4. Meanwhile, the theoretical distribution of s is nearly Normal (slightly skewed) with mean $\doteq 7.9313$ and standard deviation $\doteq 1.0458$; about 99.7% of the time, s will be between 4.8 and 11.1.

Means		Standard deviations	
22	568	5	6
23		6	
23	89	6	66899
24	02	7	3
24	89	7	
25	3	8	113
25	6799	8	789
26	124	9	000
26	59	9	556
27	4	10	2

Note: *If we take a sample of size n from a Normal distribution and compute the sample standard deviation S, then $(S/\sigma)\sqrt{n-1}$ has a "chi" distribution with $n-1$ degrees of freedom (which looks like a Normal distribution when n is reasonably large). You can learn all you would want to know—and more—about this distribution on the Web (for example, at Wikipedia). One implication of this is that "on the average," s underestimates σ; specifically, the mean of S is $\sigma \left(\frac{\sqrt{2}\,\Gamma(n/2)}{\sqrt{n-1}\,\Gamma(n/2-1/2)} \right)$. The factor in parentheses is always less than 1, but approaches 1 as n approaches infinity. The proof of this fact is left as an exercise—for the instructor, not for the average student!*

Chapter 2 Solutions

2.1. The cases are students.

2.3. With this change, the cases are cups of Mocha Frappuccino (as before). The variables (both quantitative) are size and price.

2.5. (a) The spreadsheet should look like the image on the right (especially if students use the data file from the companion CD). **(b)** There are 10 cases. **(c)** The image on the right shows the column headings used on the companion CD; some students may create their own spreadsheets and use slightly different headings. (The values of the variables should be the same.) **(d)** The variables in the second and third columns ("Bots" and "SpamsPerDay") are quantitative.

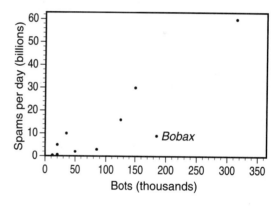

2.7. (a) The scatterplot is on the right. **(b)** Bobax is the second point from the right. (Bobax has the second-highest bot count with 185 thousand, but is relatively low in spam messages at 9 billion per day.)

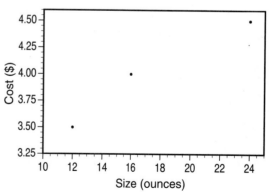

2.9. Size seems to be the most reasonable choice for explanatory variable because it seems nearly certain that Starbucks first decided which sizes to offer, then determined the appropriate price for each size (rather than vice versa). The scatterplot shows a positive association between size and price.

21

2.11. The new points (marked with a different symbol) are far away from the others, but fall roughly in the same line, so the relationship is essentially unchanged: It is still strong, linear, and positive.

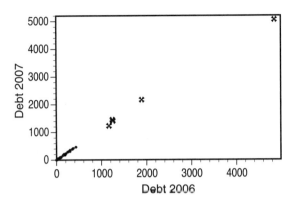

2.13. (a) A boxplot summarizes the distribution of one variable. (Two [or more] boxplots can be used to compare two [or more] distributions, but that does not allow us to examine the relationship between those variables.) **(b)** This is only correct if there is an explanatory/response relationship. Otherwise, the choice of which variable goes on which axis might be somewhat arbitrary. **(c)** High values go with high values, and low values go with low values. (Of course, those statements are generalizations; there can be exceptions.)

2.15. (a) Below, left. **(b)** Adding up the numbers in the first column of the table gives 46,994 thousand (that is, about 47 million) uninsured; divide each number in the second column by this amount. (Express answers as percents; that is, multiply by 100.) **(c)** Below, right. (The title on the vertical axis is somewhat wordy, but that is needed to distinguish between this graph and the one in the solution to the next exercise.) **(d)** The plots differ only in the vertical scale. **(e)** The uninsured are found in similar numbers for the five lowest age groups (with slightly more in those aged 25–34 and 45–64), and fewer among those over 65.

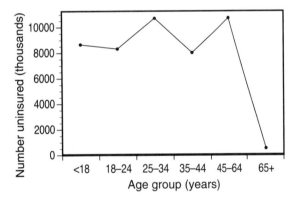
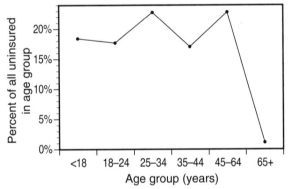

2.17. The percents in Exercise 2.15 show what fraction of the uninsured fall in each age group. The percents in Exercise 2.16 show what fraction of each age group is uninsured.
   Note: *When looking at fractions and percents, encourage students to focus on the "whole"—that is, what does the denominator represent? For all the fractions in Exercise 2.15, the "whole" is the group of all uninsured people. For Exercise 2.16, the "whole" for each fraction is the total number of people in the corresponding age group.*

2.19. (a) The scatterplot is on the right.
(b) There is a moderate positive linear relationship. (There is some suggestion of a curve, but a line seems to be a reasonable approximation given the amount of scatter.)

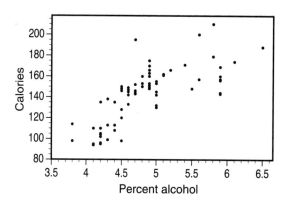

2.21. There is a moderate positive linear relationship; the relationship for all countries is less linear because of the wide range in life expectancy among countries with low Internet use.

2.23. (a) "Month" (the passage of time) explains changes in temperature (not vice versa).
(b) Temperature increases linearly with time (about 10 degrees per month); the relationship is strong.

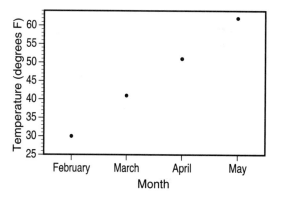

2.25. (a) The second test happens before the final exam, so that score should be viewed as explanatory. **(b)** The scatterplot shows a weak positive association. **(c)** Students' study habits are more established by the middle of the term.

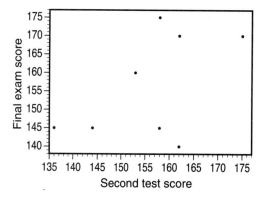

2.27. (a) Age is explanatory; weight is the response variable. **(b)** Explore the relationship; there is no reason to view one or the other as explanatory. **(c)** Number of bedrooms is explanatory; price is the response variable. **(d)** Amount of sugar is explanatory; sweetness is the response variable. **(e)** Explore the relationship.

2.29. (a) In general, we expect more intelligent children to be better readers and less intelligent children to be weaker. The plot does show this positive association. **(b)** The four points are for children who have moderate IQs but poor reading scores. **(c)** The rest of the scatterplot is roughly linear but quite weak (there would be a lot of variation about any line we draw through the scatterplot).

2.31. (a) If we used the number of males return-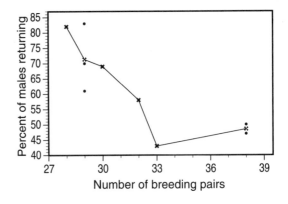
ing, then we might not see the relationship
because areas with many breeding pairs
would correspondingly have more males that
might potentially return. (In the given numbers, the number of breeding pairs varies only
from 28 to 38, but considering hypothetical
data with 10 and 100 breeding pairs makes
more apparent the reason for using percents
rather than counts.) **(b)** Scatterplot on the
right. Mean responses are shown as crosses;
the mean responses with 29 and 38 breeding pairs are (respectively) 71.3333% and 48.5%
males returning. **(c)** The scatterplot does show the negative association we would expect if
the theory were correct.

2.33. The scatterplot shows a fairly strong,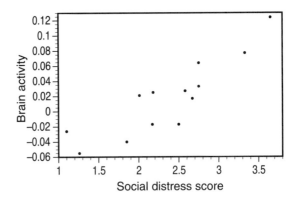
positive, linear association. There are no
particular outliers; each variable has low
and high values, but those points do not
deviate from the pattern of the rest. Social
exclusion does appear to trigger a pain response.

2.35. (a) Women are marked with filled cir-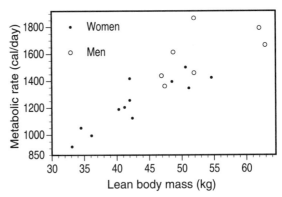
cles, men with open circles. **(b)** The asso-
ciation is linear and positive. The women's
points show a stronger association. As a
group, males typically have larger values
for both variables.

2.37. (a) Both men (filled circles) and women (open circles) show fairly steady improvement. Women have made more rapid progress, but their progress seems to have slowed (the record has not changed since 1993), while men's records may be dropping more rapidly in recent years. **(b)** The data support the first claim but do not seem to support the second.

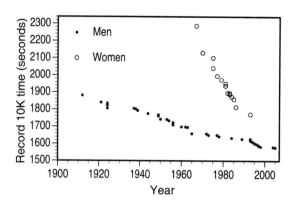

2.39. (a) The correlation is $r \doteq 0.8839$ (again). **(b)** They are equal. **(c)** Units do not affect correlation.

2.41. In both these cases, the points in a scatterplot would fall exactly on a positively sloped line, so both have correlation $r = 1$. **(a)** With $x =$ the price of a brand-name product, and $y =$ the store-brand price, the prices satisfy $y = 0.9x$. **(b)** The prices satisfy $y = x - 1$.

2.43. Software gives $r \doteq 0.2873$. (This is consistent with the not-very-strong association visible in the plot.)

2.45. The correlation is $r \doteq 0.6701$, but we note that because this relationship is not linear, r is not really appropriate for this situation.

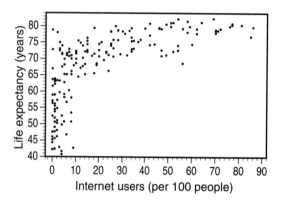

2.47. (a) $r \doteq 0.5194$. **(b)** The first-test/final-exam correlation will be lower, because the relationship is weaker. (See the next solution for confirmation.)

2.49. Such a point should be at the lower left part of the scatterplot. Because it tends to strengthen the relationship, the correlation increases.

　　Note: *In this case, r was positive, so strengthening the relationship means r gets* larger. *If r had been negative, strengthening the relationship would have* decreased *r (toward −1).*

2.51. The correlations are listed on the right; these support the observation from the solution to Exercise 2.34 that the value/debt relationship is by far the strongest.

Value and revenue $\quad r_1 \doteq -0.3228$
Value and debt $\qquad\quad r_2 \doteq 0.9858$
Value and income $\qquad r_3 \doteq 0.7177$

2.53. **(a)** The scatterplot shows a moderate positive association, so *r* should be positive, but not close to 1. **(b)** The
correlation is $r \doteq 0.5653$. **(c)** *r* would not change if all the
men were six inches shorter. A positive correlation does not
tell us that the men were generally taller than the women;
instead it indicates that women who are taller (shorter) than
the average woman tend to date men who are also taller
(shorter) than the average man. **(d)** *r* would not change because it is unaffected by units. **(e)** *r* would be 1 because the
points of the scatterplot would fall exactly on a positively
sloped line (with no scatter).

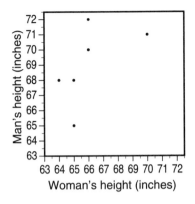

2.55. **(a)** As two points determine a line, the correlation is always either -1 or 1. **(b)** Sketches
will vary; an example is shown as the first graph below. Note that the scatterplot must be
positively sloped, but *r* is affected only by the scatter about a line drawn through the data
points, not by the steepness of the slope. **(c)** The first nine points cannot be spread from the
top to the bottom of the graph because in such a case the correlation cannot exceed about
0.66 (based on empirical evidence—that is, from a reasonable amount of playing around
with the applet). One possibility is shown as the second graph below. **(d)** To have $r \doteq 0.8$,
the curve must be higher at the right than at the left. One possibility is shown as the third
graph below.

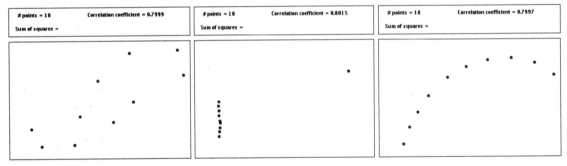

2.57. (Scatterplot not shown.) If the husband's age is *y* and the wife's *x*, the linear relationship
$y = x + 2$ would hold, and hence $r = 1$ (because the slope is positive).

2.59. The person who wrote the article interpreted a correlation close to 0 as if it were a
correlation close to -1 (implying a negative association between teaching ability and
research productivity). Professor McDaniel's findings mean there is little linear association
between research and teaching—for example, knowing that a professor is a good researcher
gives little information about whether she is a good or bad teacher.

    Note: *Students often think that "negative association" and "no association" mean the
same thing. This exercise provides a good illustration of the difference between these terms.*

2.61. Both relationships (scatterplots below) are somewhat linear. The GPA/IQ scatterplot ($r \doteq 0.6337$) shows a stronger association than GPA/self-concept ($r \doteq 0.5418$). The two students with the lowest GPAs stand out in both plots; a few others stand out in at least one plot. Generally speaking, removing these points raises r (because the remaining points look more linear). An exception: Removing the lower-left point in the self-concept plot decreases r because the relative scatter of the remaining points is greater.

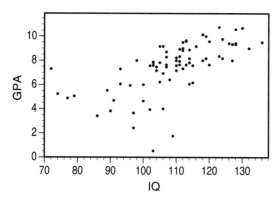 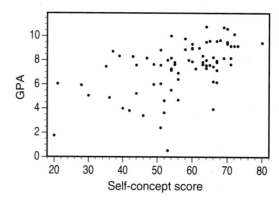

2.63. The estimated fat gain is $3.505 - 0.00344 \times 600 \doteq 1.441$ kg.

2.65. The table on the right shows the values of r^2 (expressed as percents). We observe that (i) the fraction of variation explained depends only on the magnitude (absolute value) of r, not its sign, and (ii) the fraction of explained variation drops off drastically as $|r|$ moves away from 1.

r	-0.9	-0.5	-0.3	0	0.3	0.5	0.9
r^2	81%	25%	9%	0%	9%	25%	81%

2.67. Residuals (found with software) are given in the table on the right. Los Angeles is the best; it has nearly 6000 acres more than the regression line predicts. Chicago, which falls almost 7300 acres short of the regression prediction, is the worst of this group.

Los Angeles	5994.85
Washington, D.C.	2763.75
Minneapolis	2107.59
Philadelphia	169.42
Oakland	27.91
Boston	20.96
San Francisco	-75.78
Baltimore	-131.55
New York	-282.99
Long Beach	-1181.70
Miami	-2129.21
Chicago	-7283.26

2.69. For Baltimore, for example, this rate is $\frac{5091}{651} \doteq 7.82$. The complete table is shown on the following page, left. Note that population is in thousands, so these are in units of acres per 1000 people. **(a)** Scatterplot on the following page, right. **(b)** The association is much less linear than in the scatterplot for Exercise 2.66. **(c)** The regression equation is $\hat{y} = 8.739 - 0.000424x$. **(d)** Regression on population explains only $r^2 \doteq 8.7\%$ of the variation in open space per person.

Baltimore	7.82
Boston	8.26
Chicago	4.02
Long Beach	6.25
Los Angeles	8.07
Miami	3.67
Minneapolis	14.87
New York	6.23
Oakland	9.30
Philadelphia	7.04
San Francisco	7.61
Washington, D.C.	13.12

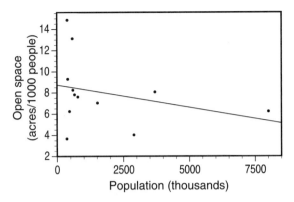

2.71. The regression equation for predicting carbohydrates from alcohol content is
$\hat{y} = 3.379 + 1.6155x$.

   Note: *As we would guess from the scatterplot, and from the correlation r $\doteq$ 0.2873 found in Exercise 2.43, this is not a very reliable prediction; it only explains r^2 $\doteq$ 8.3% of the variation in carbohydrates.*

2.73. (a) To three decimal places, the correlations are all approximately 0.816 (for set D, r actually rounds to 0.817), and the regression lines are all approximately $\hat{y} = 3.000 + 0.500x$. For all four sets, we predict $\hat{y} \doteq 8$ when $x = 10$. **(b)** Scatterplots below. **(c)** For Set A, the use of the regression line seems to be reasonable—the data do seem to have a moderate linear association (albeit with a fair amount of scatter). For Set B, there is an obvious *non*-linear relationship; we should fit a parabola or other curve. For Set C, the point (13, 12.74) deviates from the (highly linear) pattern of the other points; if we can exclude it, the (new) regression formula would be very useful for prediction. For Set D, the data point with $x = 19$ is a very influential point—the other points alone give no indication of slope for the line. Seeing how widely scattered the y coordinates of the other points are, we cannot place too much faith in the y coordinate of the influential point; thus, we cannot depend on the slope of the line, so we cannot depend on the estimate when $x = 10$. (We also have no evidence as to whether or not a line is an appropriate model for this relationship.)

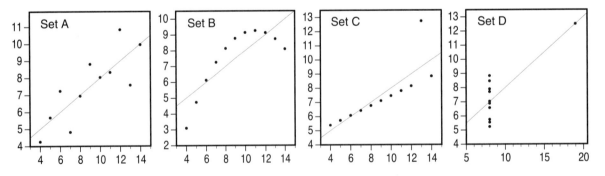

2.75. (a) The scatterplot (following page, left) suggests a fairly strong positive linear relationship. **(b)** The regression equation is $\hat{y} = 1.470 + 1.4431x$. **(c)** The residual plot is on the following page, right. The new point's residual is positive; the other residuals decrease as x increases. **(d)** The regression explains $r^2 \doteq 71.1\%$ of the variation in y. **(e)** The new point makes the relationship stronger, but its location has a large impact on the regression equation—both the slope and intercept changed substantially.

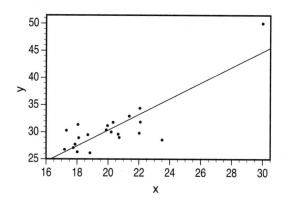 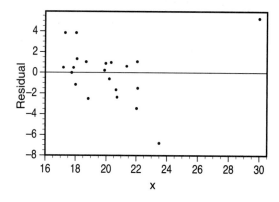

2.77. (a) When $x = 5$, $y = 12 + 6 \times 5 = 42$. **(b)** y increases by 6. (The change in y corresponding to a unit increase in x is the slope of this line.) **(c)** The intercept of this equation is 12.

2.79. See also the solution to Exercise 2.33.

    (a) The regression equation is

$$\hat{y} = 0.06078x - 0.1261$$

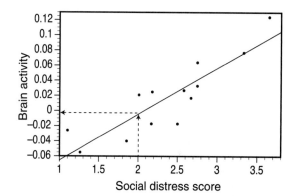

    (b) Based on the "up-and-over" method, most students will probably estimate that $\hat{y} \doteq 0$; the regression formula gives $\hat{y} = -0.0045$. **(c)** The correlation is $r \doteq 0.8782$, so the line explains $r^2 \doteq 77\%$ of the variation in brain activity.

2.81. The means and standard deviations are $\bar{x} = 95$ min, $\bar{y} \doteq 12.6611$ cm, $s_x \doteq 53.3854$ min, and $s_y \doteq 8.4967$ cm; the correlation is $r \doteq 0.9958$.

    For predicting length from time, the slope and intercept are $b_1 = r\, s_y/s_x \doteq 0.158$ cm/min and $a_1 = \bar{y} - b_1\bar{x} \doteq -2.39$ cm, giving the equation $\hat{y} = -2.39 + 0.158x$ (as in Exercise 2.80).

    For predicting time from length, the slope and intercept are $b_2 = r\, s_x/s_y \doteq 6.26$ min/cm and $a_2 = \bar{x} - b_2\bar{y} \doteq 15.79$ min, giving the equation $\hat{x} = 15.79 + 6.26y$.

2.83. The correlation of IQ with GPA is $r_1 \doteq 0.634$; for self-concept and GPA, $r_2 \doteq 0.542$. IQ does a slightly better job; it explains about $r_1^2 \doteq 40.2\%$ of the variation in GPA, while self-concept explains about $r_2^2 \doteq 29.4\%$ of the variation.

2.85. We have slope $b_1 = r\, s_y/s_x$ and intercept $b_0 = \bar{y} - b_1\bar{x}$, and $\hat{y} = b_0 + b_1x$, so when $x = \bar{x}$, $\hat{y} = b_0 + b_1\bar{x} = (\bar{y} - b_1\bar{x}) + b_1\bar{x} = \bar{y}$. (Note that the value of the slope does not actually matter.)

2.87. $r = \sqrt{0.16} = 0.40$ (high attendance goes with high grades, so r must be positive).

2.89. (a) The scatterplot (below, left) includes the regression line $\hat{y} = 10.27 - 0.5382x$. **(b)** The scatterplot looks more linear than Figure 2.5, but a line may not be appropriate for all values of log unemployment. **(c)** In the residual plot (below, right), we see that there are more negative residuals on the left and right, with more positive residuals in the middle.

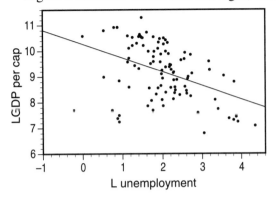
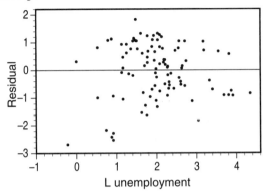

2.91. Correlations were found in Exercise 2.51. Shown below are the three scatterplots, with regression lines; the equations for those lines are given on the right. The best regression line is clearly the one based on debt.

Explanatory variable	Equation	r^2
Income	$341.54 + 3.5760x$	51.5%
Debt	$16.02 + 2.8960x$	97.2%
Revenue	$425.17 - 1.6822x$	10.4%

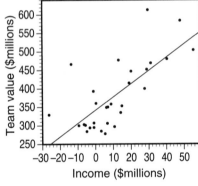
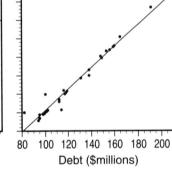
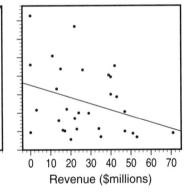

2.93. The sum of the residuals is 0.01.

2.95. (a) A high correlation means strong association, not causation. **(b)** Outliers in the y direction (and some other data points) will have large residuals. **(c)** It is not extrapolation if $1 \leq x \leq 5$.

2.97. A reasonable explanation is that the cause-and-effect relationship goes in the other direction: Doing well makes students or workers feel good about themselves, rather than vice versa.

2.99. The explanatory and response variables were "consumption of herbal tea" and "cheerfulness/health." The most important lurking variable is social interaction; many of the nursing home residents may have been lonely before the students started visiting.

2.101. (a) The plot (below, left) is curved (low at the beginning and end of the year, high in the middle). **(b)** The regression line is $\hat{y} = 39.392 + 1.4832x$. It does not fit well because a line is poor summary of this relationship. **(c)** Residuals are negative for January through March and October through December (when actual temperature is less than predicted temperature), and positive from April to September (when it is warmer than predicted). **(d)** A similar pattern would be expected in any city that is subject to seasonal temperature variation. **(e)** Seasons in the Southern Hemisphere are reversed, so temperature would be cooler in the middle of the year.

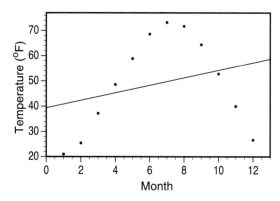 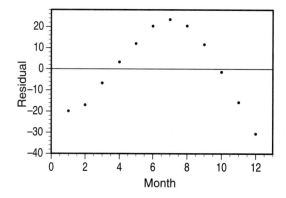

2.103. With individual children, the correlation would be smaller (closer to 0) because the additional variation of data from individuals would increase the "scatter" on the scatterplot, thus decreasing the strength of the relationship.

2.105. For example, a student who in the past might have received a grade of B (and a lower SAT score) now receives an A (but has a lower SAT score than an A student in the past). While this is a bit of an oversimplification, this means that today's A students are yesterday's A and B students, today's B students are yesterday's C students, and so on. Because of the grade inflation, we are not comparing students with equal abilities in the past and today.

2.107. The correlation between BMR and fat gain is $r \doteq 0.08795$; the slope of the regression line is $b = 0.000811$ kg/cal. These both show that BMR is less useful for predicting fat gain. The small correlation suggests a very weak linear relationship (explaining less than 1% of the variation in fat gain). The small slope means that changes in BMR have very little impact on fat gain; for example, increasing BMR by 100 calories changes fat gain by only 0.08 kg.

2.109. (a) Scatterplot on the right. (b) The plot shows a strong positive linear relationship. (c) The regression equation is $\hat{y} = 20.40 + 0.7194x$. (d) Hernandez's point is in the lower left—a logical place for the eventual champion.

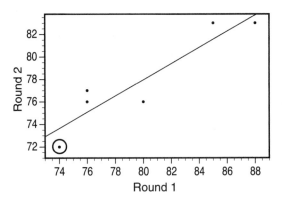

2.111. (a) Drawing the "best line" by eye is a very inaccurate process; few people choose the best line (although you can get better at it with practice). (b) Most people tend to overestimate the slope for a scatterplot with $r \doteq 0.7$; that is, most students will find that the least-squares line is less steep than the one they draw.

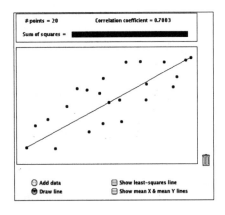 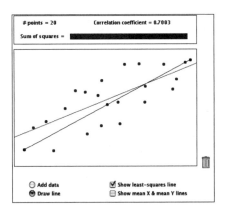

2.113. The plot shown is a very simplified (and not very realistic) example. Filled circles are economists in business; open circles are teaching economists. The plot should show positive association when either set of circles is viewed separately and should show a large number of bachelor's degree economists in business and graduate degree economists in academia.

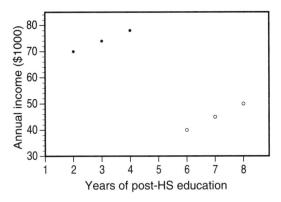

2.115. There are 1684 female binge drinkers in the table; 8232 female students are not binge drinkers.

2.117. Divide the number of non-bingeing females by the total number of students:

$$\frac{8232}{17,096} \doteq 0.482$$

2.119. This is a conditional distribution; take the number of bingeing males divided by the total number of males: $\frac{1630}{7180} \doteq 0.227$.

2.121. (a) There are $151 + 148 = 299$ "high exercisers," of which $\frac{151}{299} \doteq 50.5\%$ get enough sleep and 49.5% (the rest) do not. **(b)** There are $115 + 242 = 357$ "low exercisers," of which $\frac{115}{357} \doteq 32.2\%$ get enough sleep and 67.8% (the rest) do not. **(c)** Those who exercise more than the median are more likely to get enough sleep.

    Note: *This question is asking for the conditional distribution of sleep within each exercise group. Exercise 2.122 asks for the conditional distribution of exercise within each sleep group.*

2.123. $\frac{63}{2100} = 3.0\%$ of Hospital A's patients died, compared with $\frac{16}{800} = 2.0\%$ at Hospital B.

2.125. (a) There are about 3,388,000 full-time college students aged 15 to 19. (Note that numbers are in thousands.) **(b)** The joint distribution is found by dividing each number in the table by 16,388 (the total of all the numbers). These proportions are given in italics in the table on the right. For example, $\frac{3388}{16,388} \doteq 0.2067$, meaning that about 20.7% of all college students are full-time and aged 15 to 19. **(c)** The marginal distribution of age is found by dividing the *row* totals by 16,388; they are in the right margin of the table and the graph on the left below. For example, $\frac{3777}{16,388} \doteq 0.2305$,

	FT	PT	
15–19	3388	389	3777
	0.2067	*0.0237*	*0.2305*
20–24	5238	1164	6402
	0.3196	*0.0710*	*0.3907*
25–34	1703	1699	3402
	0.1039	*0.1037*	*0.2076*
35+	762	2045	2807
	0.0465	*0.1248*	*0.1713*
	11091	5297	16388
	0.6768	*0.3232*	

meaning that about 23% of all college students are aged 15 to 19. **(d)** The marginal distribution of status is found by dividing the *column* totals by 16,388; they are in the bottom margin of the table and the graph on the right below. For example, $\frac{11,091}{16,388} \doteq 0.6768$, meaning that about 67.7% of all college students are full-time.

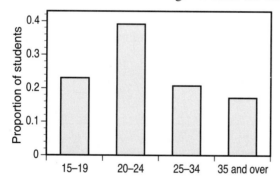
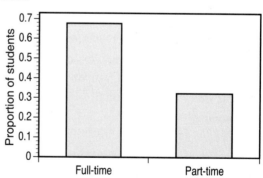

2.127. Refer to the counts in the solution to Exercise 2.125. For each status category, the conditional distribution of age is found by dividing the counts in that column by that column total. For example, $\frac{3388}{11,091} \doteq 0.3055$, $\frac{5238}{11,091} \doteq 0.4723$, etc., meaning that of all full-time college students, about 30.55% are aged 15 to 19, 47.23% are 20 to 24, and so on. Note that each set of four numbers should add up to 1 (except for rounding error). Graphical presentations may vary; one possibility is shown below. We see that full-time students are dominated by younger ages, while part-time students are more likely to be older. (This is essentially the same observation made in the previous exercise, seen from a different viewpoint.)

	FT	PT
15–19	0.3055	0.0734
20–24	0.4723	0.2197
25–34	0.1535	0.3207
35+	0.0687	0.3861

2.129. To construct such a table, we can start by choosing values for the row and column sums $r1, r2, r3, c1, c2, c3$, as well as the grand total N. Note that $N = r1 + r2 + r3 = c1 + c2 + c3$, so we only have five choices to make. Then, find each count $a, b, c, d, e, f, g, h, i$ by taking the corresponding *row* total, times the corresponding *column* total, divided by the *grand* total. For example, $a = r1 \times c1/N$ and $f = r2 \times c3/N$. Of course, these counts should be whole numbers, so it may be necessary to make adjustments in the row and column totals to meet this requirement.

a	b	c	r1
d	e	f	r2
g	h	i	r3
c1	c2	c3	N

The simplest such table would have all nine counts ($a, b, c, d, e, f, g, h, i$) equal to one another.

2.131. (a) Use column percents, e.g., $\frac{68}{225} \doteq 30.22\%$ of females are in administration. See table and graph below. The biggest difference between women and men is in Administration: A higher percentage of women chose this major. Meanwhile, a greater proportion of men chose other fields, especially Finance. **(b)** There were 386 responses; $\frac{336}{722} \doteq 46.5\%$ did not respond.

	Female	Male	Overall
Accting.	30.22%	34.78%	32.12%
Admin.	40.44%	24.84%	33.94%
Econ.	2.22%	3.70%	2.85%
Fin.	27.11%	36.65%	31.09%

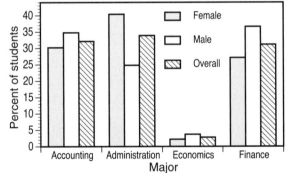

2.133. This is a case of confounding: The association between dietary iron and anemia is difficult to detect because malaria and helminths also affect iron levels in the
body.

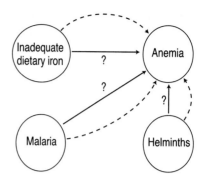

2.135. Responses will vary. For example, students
who choose the online course might have more self-
motivation or better computer skills. A diagram is shown
on the right; the generic "Student characteristics" might
be replaced with something more specific.

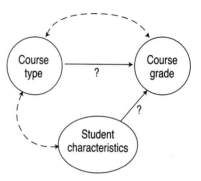

2.137. No; self-confidence and improving fitness could be a common response to some other
personality trait, or high self-confidence could make a person more likely to join the
exercise program.

2.139. Two possibilities are that they might perform better simply because this is their second
attempt or because they feel better prepared as a result of taking the course (whether or not
they really *are* better prepared).

2.141. Patients suffering from more serious illnesses are
more likely to go to larger hospitals (which may have
more or better facilities) for treatment. They are also
likely to require more time to recuperate afterwards.

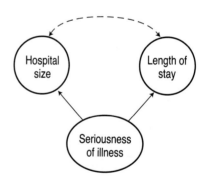

2.143. In this case, there may be a causative effect, but in the direction opposite to the one
suggested: People who are overweight are more likely to be on diets and so choose artificial
sweeteners over sugar. (Also, heavier people are at a higher risk to develop diabetes; if they
do, they are likely to switch to artificial sweeteners.)

2.145. This is an observational study—students choose their "treatment" (to take or not take
the refresher sessions).

2.149. (a) The residuals are positive at the beginning and end, and negative in the middle. **(b)** The behavior of the residuals agrees with the curved relationship seen in Figure 2.30.

2.151. (a) The regression equation for predicting salary from year is $\hat{y} = 41.253 + 3.9331x$; for $x = 25$, the predicted salary is $\hat{y} \doteq 139.58$ thousand dollars, or about \$139,600. **(b)** The log-salary regression equation is $\hat{y} = 3.8675 + 0.04832x$. With $x = 25$, we have $\hat{y} \doteq 5.0754$, so the predicted salary is $e^{\hat{y}} \doteq 160.036$, or about \$160,040. **(c)** Although both predictions involve extrapolation, the second is more reliable, because it is based on a linear fit to a linear relationship. **(d)** Interpreting relationships without a plot is risky. **(e)** Student summaries will vary, but should include comments about the importance of looking at plots, and the risks of extrapolation.

2.153. (a) The regression equation is $\hat{y} = 5403 + 0.9816x$. **(b)** The residual plot on the right reveals no causes for concern; in particular, there are no clear outliers or influential observations.

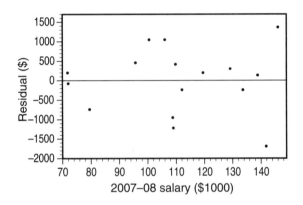

2.155. A school that accepts weaker students but graduates a higher-than-expected number of them would have a positive residual, while a school with a stronger incoming class but a lower-than-expected graduation rate would have a negative residual. It seems reasonable to measure school quality by how much benefit students receive from attending the school.

2.157. (a) The scatterplot shows a moderate positive association. **(b)** The regression line $\hat{y} = 11.00 + 0.9344x$ fits the overall trend. (However, there is some hint of a curve in the plot, because most of the points in the middle lie below the line.) **(c)** For example, a state whose point falls above the line has a higher percent of college graduates than we would expect based on the percent who eat 5 servings of fruits and vegetables. **(d)** No; association is not evidence of causation.

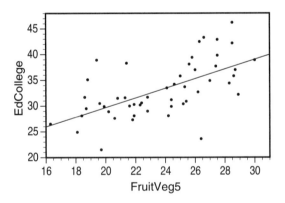

2.161. These results support the idea (the slope is negative, so variation decreases with increasing diversity), but the relationship is only moderately strong ($r^2 = 0.34$, so diversity only explains 34% of the variation in population variation).

   Note: *That last parenthetical comment is awkward and perhaps confusing, but is consistent with similar statements interpreting r^2.*

2.163. On the right is a scatterplot of MOR against MOE, showing a moderate linear positive association. The regression equation is $\hat{y} = 2653 + 0.004742x$; this regression explains $r^2 = 0.6217 \doteq 62\%$ of the variation in MOR. So, we can use MOE to get fairly good (though not perfect) predictions of MOR.

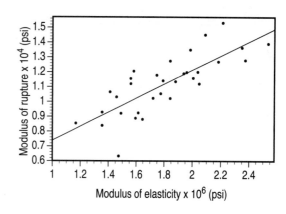

2.165. **(a)** The scatterplot is on the right. **(b)** The regression equation is $\hat{y} = 1.2027 + 0.3275x$. As we see from the scatterplot, the relationship is not too strong; the correlation ($r = 0.4916$, $r^2 = 0.2417$) confirms this.

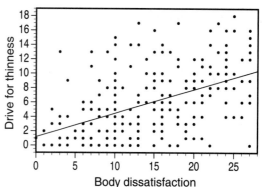

2.167. **(a)** Shown below are plots of count against time, and residuals against time for the regression, which gives the formula $\hat{y} = 259.58 - 19.464x$. Both plots suggest a curved relationship rather than a linear one. **(b)** With natural logarithms, the regression equation is $\hat{y} = 5.9732 - 0.2184x$; with common logarithms, $\hat{y} = 2.5941 - 0.09486x$. The second pair of plots (following page) show the (natural) logarithm of the counts against time, suggesting a fairly linear relationship, and the residuals against time, which shows no systematic pattern. (If common logarithms are used instead of natural logs, the plots will look the same, except the vertical scales will be different.) The correlations confirm the increased linearity of the log plot: $r^2 = 0.8234$ for the original data, $r^2 = 0.9884$ for the log-data.

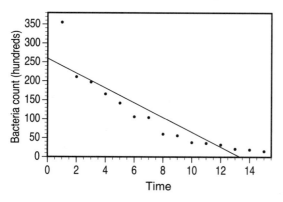

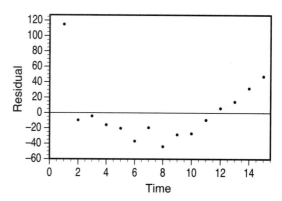

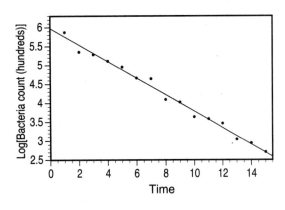

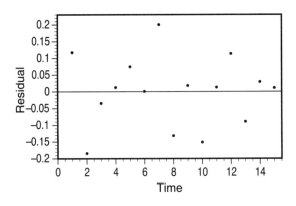

2.169. Number of firefighters and amount of damage both increase with the seriousness of the fire (i.e., they are common responses to the fire's seriousness).

2.171. Different graphical presentations are possible; one is shown below. More women perform volunteer work; the notably higher percentage of women who are "strictly voluntary" participants accounts for the difference. (The "court-ordered" and "other" percentages are similar for men and women.)

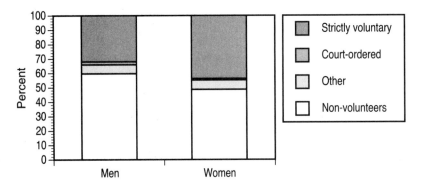

2.173. (a) At right. **(b)** $\frac{490}{800} = 61.25\%$ of male applicants are admitted, while only $\frac{400}{700} \doteq 57.14\%$ of females are admitted.

	Admit	Deny
Male	490	310
Female	400	300

(c) $\frac{400}{600} \doteq 66.67\%$ of male business school applicants are admitted; for females, this rate is the same: $\frac{200}{300} \doteq 66.67\%$. In the law school, $\frac{90}{200} = 45\%$ of males are admitted, compared to $\frac{200}{400} = 50\%$ of females. **(d)** A majority (6/7) of male applicants apply to the business school, which admits $\frac{400+200}{600+300} \doteq 66.67\%$ of all applicants. Meanwhile, a majority (3/5) of women apply to the law school, which admits only $\frac{90+200}{200+400} \doteq 48.33\%$ of its applicants.

2.175. If we ignore the "year" classification, we see that Department A teaches 32 small classes out of 52, or about 61.54%, while Department B teaches 42 small classes out of 106, or about 39.62%. (These agree with the dean's numbers.)

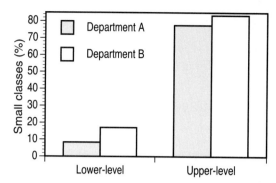

For the report to the dean, students may analyze the numbers in a variety of ways, some valid and some not. The key observations are: (i) When considering only first- and second-year classes, A has fewer small classes ($\frac{1}{12} \doteq 8.33\%$) than B ($\frac{12}{70} \doteq 17.14\%$). Likewise, when considering only upper-level classes, A has $\frac{31}{40} = 77.5\%$ and B has $\frac{30}{36} \doteq 83.33\%$ small classes. The graph on the right illustrates this. These numbers are given in the back of the text, so most students should include this in their analysis! (ii) $\frac{40}{52} \doteq 77.78\%$ of A's classes are upper-level courses, compared to $\frac{36}{106} \doteq 33.96\%$ of B's classes.

2.177. (a) First we must find the counts in each cell of the two-way table. For example, there were about $(0.172)(5619) \doteq 966$ Division I athletes who admitted to wagering. These counts are shown in the Minitab output on the right, where we see that $X^2 \doteq 76.7$, df = 2, and $P < 0.0001$. There is very strong evidence that the percentage of athletes who admit to wagering differs by division.

Minitab output

	Div1	Div2	Div3	Total
1	966	621	998	2585
	1146.87	603.54	834.59	
2	4653	2336	3091	10080
	4472.13	2353.46	3254.41	
Total	5619	2957	4089	12665

```
ChiSq = 28.525 +  0.505 + 31.996 +
         7.315 +  0.130 +  8.205 = 76.675
df = 2, p = 0.000
```

(b) Even with much smaller numbers of students (say, 1000 from each division), P is still very small. Presumably, the estimated numbers are reliable enough that we would not expect the true counts to be less than 1000, so we need not be concerned about the fact that we had to estimate the sample sizes. **(c)** If the reported proportions are wrong, then our conclusions may be suspect—especially if it is the case that athletes in some division were more likely to say they had not wagered when they had. **(d)** It is difficult to predict exactly how this might affect the results: Lack of independence could cause the estimated percents to be too large, or too small, if our sample included several athletes from teams which have (or do not have) a "gambling culture."

Chapter 3 Solutions

3.1. Any group of friends is unlikely to include a representative cross section of all students.

3.3. A hard-core runner (and her friends) are not representative of all young people.

3.5. For example, who owns the Web site? Do they have data to back up this statement, and if so, what was the source of that data?

3.7. This is an experiment: Each subject is assigned to a treatment group (presumably at random, although the description does not tell us this is the case). The explanatory variable is the drug received, the response variables are adverse events, as well as some reaction. (The nature of that reaction is not specified in the exercise, but they apparently collected some information to indicate which subjects had an "effective response" to the vaccine.)

3.9. This is an experiment: Each subject is (presumably randomly) assigned to a group, each with its own treatment (computer animation or reading the textbook). The explanatory variable is the teaching method, and the response variable is the change in each student's test score.

3.11. The experimental units are food samples, the treatment is exposure to different levels of radiation, and the response variable is the amount of lipid oxidation. Note that in a study with only one factor—like this one—the treatments and factor levels are essentially the same thing: The factor is varying radiation exposure, with nine levels.
   It is hard to say how much this will generalize; it seems likely that different lipids react to radiation differently.

3.13. Those who volunteer to use the software may be better students (or worse). Even if we cannot decide the direction of the bias (better or worse), the lack of random allocation means that the conclusions we can draw from this study are limited at best.

3.15. Because there are nine levels, this diagram is rather large (and repetitive), so only the top three branches are shown.

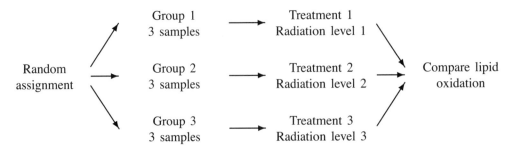

3.17. (a) Shopping patterns may differ on Friday and Saturday, which would make it hard to determine the true effect of each promotion. (That is, the effect of the promotion would be confounded with the effect of the day.) To correct this, we could offer one promotion on a

Friday, and the other on the following Friday. (Or, we could do as described in the exercise, and then on the next weekend, swap the order of the offers.) **(b)** Responses may vary in different states. To control for this, we could launch both campaigns in (separate) parts of the same state or states. **(c)** A control is needed for comparison; if we simply compare this year's yield to last year's yield, we will not know how much of the difference can be attributed to changes in the economy. We should compare the new strategy's yield with another investment portfolio using the old strategy.

> **Note:** *For part (c), this comparison might be done without actually buying or selling anything; we could simply compute how much money would have been made if we had followed the new strategy; that is, we keep a "virtual portfolio." This assumes that our buying and selling is on a small enough scale that it does not affect market prices.*

3.19. The experiment can be single-blind (those evaluating the exams should not know which teaching approach was used), but not double-blind, because the students will know which treatment (teaching method) was assigned to them.

3.21. For example, new employees should be randomly assigned to either the current program or the new one. There are many possible choices for outcome variables, such as performance on a test of the information covered in the program, or a satisfaction survey or other evaluation of the program by those who went through it.

3.23. **(a)** The factors are calcium dose, and vitamin D dose. There are 9 treatments (each calcium/vitamin D combination). **(b)** Assign 20 students to each group, with 10 of each gender. The complete diagram (including the blocking step) would have a total of 18 branches, below is a portion of that diagram, showing only three of the nine branches for each gender. **(c)** Randomization results will vary.

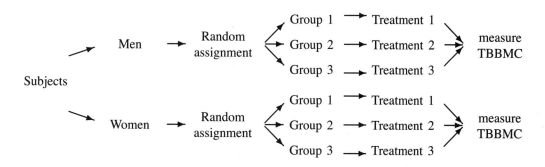

3.25. **(a)** For example, flip a coin for each customer to choose which variety (s)he will taste. To evaluate preferences, we would need to design some scale for customers to rate the coffee they tasted, and then compare the ratings. **(b)** For example, flip a coin for each customer to choose which variety (s)he will taste *first*. Ask which of the two coffees (s)he preferred. **(c)** The matched-pairs version described in part (b) is the stronger design; if each customer tastes both varieties, we only need to ask which was preferred. In part (a), we might ask customers to rate the coffee they tasted on a scale of (say) 1 to 10, but such ratings can be wildly variable (one person's "5" might be another person's "8"), which makes comparison of the two varieties more difficult.

3.27. Experimental units: pine tree seedlings. Factor: amount of light. Treatments (two): full light, or shaded to 5% of normal. Response variable: dry weight at end of study.

3.29. Subjects: adults (or registered voters) from selected households. Factors: level of identification, and offer of survey results. Treatments (six): interviewer's name with results, interviewer's name without results, university name with results, university name without results, both names with results, both names without results. Response variable: whether or not the interview is completed.

3.31. Assign nine subjects to treatment 1, then nine more to treatment 2, etc. A diagram is below; if we assign labels 01 through 36, then line 151 gives:

Group 1		Group 2		Group 3	
03 Bezawada	12 Hatfield	32 Tyner	27 Rau	05 Cheng	13 Hua
22 Mi	11 Guha	30 Tang	20 Martin	16 Leaf	25 Park
29 Shu	31 Towers	09 Daye	06 Chronopoulou	28 Saygin	19 Lu
26 Paul	21 Mehta	23 Nolan	33 Vassilev	10 Engelbrecht	04 Cetin
01 Anderson		07 Codrington		18 Lipka	

The other nine subjects are in Group 4. The names listed here assume that labels are assigned alphabetically (across the rows). See note on page 1 about using Table B.

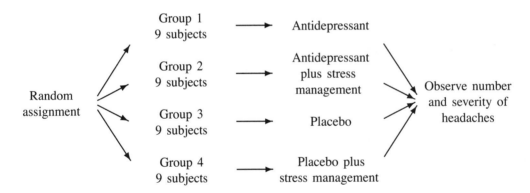

3.33. Students might envision different treatments; one possibility is to have some volunteers go through a training session, while others are given a written set of instructions, or watch a video. For the response variable(s), we need some measure of training effectiveness; perhaps we could have the volunteers analyze a sample of lake water and compare their results to some standard.

3.35. Diagram on the following page. Starting at line 160, we choose:
    16, 21, 06, 12, 02, 04 for Group 1
    14, 15, 23, 11, 09, 03 for Group 2
    07, 24, 17, 22, 01, 13 for Group 3
The rest are assigned to Group 4. See note on page 1 about using Table B.

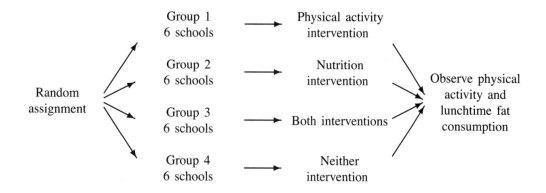

3.37. **(a)** Population = 1 to 150 , Select a sample of size 25 , click Reset and Sample .
(b) Without resetting, click Sample again. **(c)** Click Sample three more times.

3.39. Design (a) is an experiment. Because the treatment is randomly assigned, the effect of other habits would be "diluted" because they would be more-or-less equally split between the two groups. Therefore, any difference in colon health between the two groups could be attributed to the treatment (bee pollen or not).

Design (b) is an observational study. It is flawed because the women observed chose whether or not to take bee pollen; one might reasonably expect that people who choose to take bee pollen have other dietary or health habits that would differ from those who do not.

3.41. As described, there are two factors: ZIP code (three levels: none, five-digit, nine-digit) and the day on which the letter is mailed (three levels: Monday, Thursday, or Saturday) for a total of nine treatments. To control lurking variables, aside from mailing all letters to the same address, all letters should be the same size, and either printed in the same handwriting or typed. The design should also specify how many letters will be in each treatment group. Also, the letters should be sent randomly over many weeks.

3.43. Each player will be put through the sequence (100 yards, four times) twice—once with oxygen and once without, and we will observe the difference in their times on the final run. (If oxygen speeds recovery, we would expect that the oxygen-boosted time will be lower.) Randomly assign half of the players to use oxygen on the first trial, while the rest use it on the second trial. Trials should be on different days to allow ample time for full recovery.

If we label the players 01 through 20 and begin on line 140, we choose 12, 13, 04, 18, 19, 16, 02, 08, 17, 10 to be in the oxygen-first group. See note on page 1 about using Table B.

3.45. **(a)** Randomly assign half the girls to get high-calcium punch; the other half will get low-calcium punch. The response variable is not clearly described in this exercise; the best we can say is "observe how the calcium is processed." **(b)** Randomly select half of the girls to receive high-calcium punch first (and low-calcium punch later), while the other half gets low-calcium punch first (followed by high-calcium punch). For each subject, compute the difference in the response variable for each level. This is a better design because it deals with person-to-person variation; the differences in responses for 60 individuals gives more precise results than the difference in the average responses for two groups of 30 subjects. **(c)** The first five subjects are 38, 44, 18, 33, and 46. In the CR design, the first group receives high-calcium punch all summer; in the matched pairs design, they receive high-calcium punch for the first part of the summer, and then low-calcium punch in the second half.

3.47. (a) This is a block design. **(b)** The diagram might be similar to the one below (which assumes equal numbers of subjects in each group). **(c)** The results observed in this study would rarely have occurred by chance if vitamin C were ineffective.

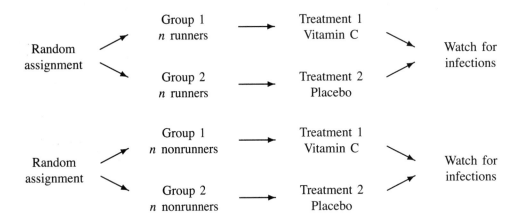

3.49. The population is all forest owners in the region. The sample is the 772 forest owners contacted. The response rate is $\frac{348}{772} \doteq 45\%$. Aside from the given information, we would like to know the sample design (and perhaps some other things).

    Note: *It would also be reasonable to consider the sample to be the 348 forest owners who returned the survey, because that is the actual group from which we "draw conclusions about the whole."*

3.51. To use Table B, number the list from 0 to 9 and choose three single digits. (One can also assign labels 01–10, but that would require two-digit numbers, and we would almost certainly end up skipping over many pairs of digits before we found three in the desired range.)

    It is worth noting that choosing an SRS is often described as "pulling names out of a hat." For long lists, it is often impractical to do this literally, but with such a small list, one really could write each song on a slip of paper and choose two slips at random.

3.53. (a) This statement confuses the ideas of population and sample. (If the entire population is found in our sample, we have a *census* rather than a sample.) **(b)** "Dihydrogen monoxide" is H_2O. Any concern about the dangers posed by water most likely means that the respondent did not know what dihydrogen monoxide was, and was too embarrassed to admit it. (Conceivably, the respondent knew the question was about water and had concerns arising from a bad experience of flood damage or near-drowning. But misunderstanding seems to be more likely.) **(c)** Honest answers to such questions are difficult to obtain even in an anonymous survey; in a public setting like this, it would be surprising if there were any raised hands (even though there are likely to be at least a few cheaters in the room).

3.55. The population is (all) local businesses. The sample is the 73 businesses that return the questionnaire, *or* the 150 businesses selected. The nonresponse rate is $51.3\% = \frac{77}{150}$.

    Note: *The definition of "sample" makes it somewhat unclear whether the sample includes all the businesses selected or only those that responded. My inclination is toward the latter (the smaller group), which is consistent with the idea that the sample is "a part of the population that we actually examine."*

3.57. Note that the numbers add to 100% down the columns; that is, 39% is the percent of Fox viewers who are Republicans, *not* the percent of Republicans who watch Fox.

    Students might display the data using a stacked bar graph like the one below, or side-by-side bars. (They could also make four pie charts, but comparing slices across pies is difficult.) The most obvious observation is that the party identification of Fox's audience is noticeably different from the other three sources.

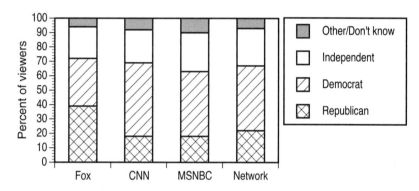

3.59. Numbering from 01 to 33 alphabetically (down the columns), we enter Table B at line 137 and choose:

    12 = Country View, 14 = Crestview, 11 = Country Squire, 16 = Fairington, 08 = Burberry
See note on page 1 about using Table B.

3.61. Population = 1 to **200** , Select a sample of size **25** , then click **Reset** and **Sample** .

3.63. One could use the labels already assigned to the blocks, but that would mean skipping a lot of four-digit combinations that do not correspond to any block. An alternative would be to drop the second digit and use labels 100–105, 200–211, and 300–325. But by far the simplest approach is to assign labels 01–44 (in numerical order by the four-digit numbers already assigned), enter the table at line 135, and select:

    39 (block 3020), 10 (2003), 07 (2000), 11 (2004), and 20 (3001)
See note on page 1 about using Table B.

3.65. The sample will vary with the starting line in Table B. The simplest method is to use the last digit of the numbers assigned to the blocks in Group 1 (that is, assign the labels 0–5), then choose one of those blocks; use the last two digits of the blocks in Group 2 (00–11) and choose two of those, and finally use the last two digits of the blocks in Group 3 (00–25) and choose three of them.

3.67. Considering the 9000 students of Exercise 3.66, each student is equally likely; specifically, each name has chance 1/45 of being selected. To see this, note that each of the first 45 has chance 1/45 because one is chosen at random. But each student in the second 45 is chosen exactly when the corresponding student in the first 45 is, so each of the second 45 also has chance 1/45. And so on.

This is not an SRS because the only possible samples have exactly one name from the first 45, one name from the second 45, and so on; that is, there are only 45 possible samples. An SRS could contain *any* 200 of the 9000 students in the population.

3.69. Assign labels 01–36 for the Climax 1 group, 01–72 for the Climax 2 group, and so on. Then beginning at line 140, choose:

12, 32, 13, 04 from the Climax 1 group and (continuing on in Table B)
51, 44, 72, 32, 18, 19, 40 from the Climax 2 group
24, 28, 23 from the Climax 3 group and
29, 12, 16, 25 from the mature secondary group

See note on page 1 about using Table B.

3.71. Each student has a 10% chance: 3 out of 30 over-21 students, and 2 of 20 under-21 students. This is not an SRS because not every group of 5 students can be chosen; the only possible samples are those with 3 older and 2 younger students.

3.73. (a) This design would omit households without telephones or with unlisted numbers. Such households would likely be made up of poor individuals (who cannot afford a phone), those who choose not to have phones, and those who do not wish to have their phone numbers published. **(b)** Those with unlisted numbers would be included in the sampling frame when a random-digit dialer is used.

3.75. The first wording brought the higher numbers in favor of a tax cut; "new government programs" has considerably less appeal than the list of specific programs given in the second wording.

3.79. The population is undergraduate college students. The sample is the 2036 students. (We assume they were randomly selected.)

3.81. The larger sample would have less sampling variability. (That is, the results would have a higher probability of being closer to the "truth.")

3.83. (a) Population: college students. Sample: 17,096 students. **(b)** Population: restaurant workers. Sample: 100 workers. **(c)** Population: longleaf pine trees. Sample: 40 trees.

3.85. **(a)** The scores will vary depending on the starting row. Note that the smallest possible mean is 5.25 (from the sample 3, 5, 6, 7) and the largest is 11.5 (from 9, 10, 12, 15). **(b)** Answers will vary. On the right is the (exact) sampling distribution, showing all possible values of the experiment (so the first rectangle is for 5.25, the next is for 5.5, etc.). Note that it looks roughly Normal; if we had taken a larger sample from a larger population, it would appear even more Normal.

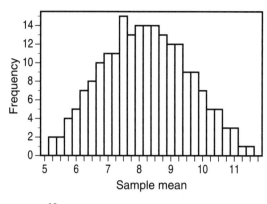

Note: *This histogram was found by considering all $\binom{10}{4} = 210$ of the possible samples. A collection of only 10 random samples will, of course, be considerably less detailed.*

3.87. **(a)** This is a multistage sample. **(b)** Attitudes in smaller countries (many of which were not surveyed) might be different. **(c)** An individual country's reported percent will typically differ from its true percent by no more than the stated margin of error. (The margins of error differ among the countries because the sample sizes were not all the same.)

   Note: *The number of countries in the world is about 195 (the exact number depends on the criteria of what constitutes a separate country). That means that about 60 countries are not represented in this survey.*

3.89. **(a)** The histogram should be centered at about 0.6 (with quite a bit of spread). For reference, the theoretical histogram is shown below on the left; student results should have a similar appearance. **(b)** The histogram should be centered at about 0.2 (with quite a bit of spread). The theoretical histogram is shown below on the right.

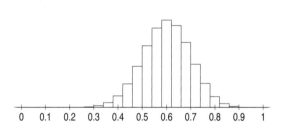 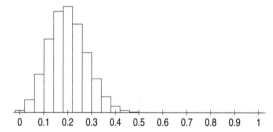

3.91. **(a)** The scores will vary depending on the starting row. Note that the smallest possible mean is 61.75 (from the sample 58, 62, 62, 65) and the largest is 77.25 (from 73, 74, 80, 82). **(b)** Answers will vary; shown on the following page are two views of the (exact) sampling distribution. The first shows all possible values of the experiment (so the first rectangle is for 61.75, the next is for 62.00, etc.); the other shows values grouped from 61 to 61.75, 62 to 62.75, etc. (which makes the histogram less bumpy). The tallest rectangle in the first picture is 8 units; in the second, the tallest is 28 units.

   Note: *These histograms were found by considering all $\binom{10}{4} = 210$ of the possible* *samples. It happens that half (105) of those samples yield a mean smaller than 69.4 and half yield a greater mean.*

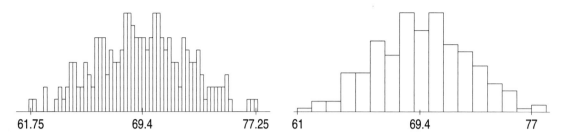

3.93. (a) Below is the population stemplot (which gives the same information as a histogram). The (population) mean GPA is $\mu \doteq 2.6352$, and the standard deviation is $\sigma \doteq 0.7794$. [Technically, we should take $\sigma \doteq 0.7777$, which comes from dividing by n rather than $n - 1$, but few (if any) students would know this, and it has little effect on the results.] **(b)** & **(c)** Results will vary; these histograms are not shown. Not every sample of size 20 could be viewed as "generally representative of the population," but most should bear at least some resemblance to the population distribution.

```
0 | 134
0 | 567889
1 | 0011233444
1 | 5566667888888888999999
2 | 000000000111111111122222222223333333333444444444
2 | 5555555555555566666666677777777777777888888888888999999
3 | 0000000000000011111111111222222222233333333333333333444444444
3 | 556666666667777778888889
4 | 0000
```

3.95. (a) Answers will vary. If, for example, eight heads are observed, then $\hat{p} = \frac{8}{20} = 0.4 = 40\%$. **(b)** Note that all the leaves in the stemplot should be either 0 or 5 since all possible $\hat{p}$-values end in 0 or 5. For comparison, here is a histogram of the sampling distribution (assuming p really is 0.5). An individual student's stemplot will probably only roughly approximate this distribution, but pooled efforts should be fairly close.

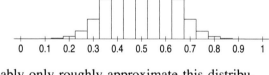

Many of the questions in Section 3.4 (Ethics), Exercises 3.96–3.117, are matters of opinion and may be better used for class discussion rather than as assigned homework. A few comments are included here.

3.97. (a) A nonscientist might raise different viewpoints and concerns from those considered by scientists. **(b)** Answers will vary.

3.103. To control for changes in the mass spectrometer over time, we should alternate between control and cancer samples.

3.105. The articles are "Facebook and academic performance: Reconciling a media sensation with data" (Josh Pasek, eian more, Eszter Hargittai), a critique of the first article called "A response to reconciling a media sensation with data" (Aryn C. Karpinski), and a response to the critique ("Some clarifications on the Facebook-GPA study and Karpinski's response") by the original authors. In case these articles are not available at the address given in the text, they might be found elsewhere with a Web search.

3.111. They cannot be anonymous because the interviews are conducted in person in the subject's home. They are certainly kept confidential.

    Note: *For more information about this survey, see the GSS Web site:*

$$\texttt{www.norc.org/GSS+Website}$$

3.121. (a) You need information about a random selection of his games, not just the ones he chooses to talk about. **(b)** These students may have chosen to sit in the front; all students should be randomly assigned to their seats.

3.123. This is an experiment because each subject is (randomly, we assume) assigned to a treatment. The explanatory variable is the price history seen by the subject (steady prices or fluctuating prices), and the response variable is the price the subject expects to pay.

3.127. The two factors are gear (three levels) and steepness of the course (number of levels not specified). Assuming there are at least three steepness levels—which seems like the smallest reasonable choice—that means at least nine treatments. Randomization should be used to determine the order in which the treatments are applied. Note that we must allow ample recovery time between trials, and it would be best to have the rider try each treatment several times.

3.129. (a) One possible population: all full-time undergraduate students in the fall term on a list provided by the registrar. **(b)** A stratified sample with 125 students from each year is one possibility. **(c)** Mailed (or emailed) questionnaires might have high nonresponse rates. Telephone interviews exclude those without phones and may mean repeated calling for those who are not home. Face-to-face interviews might be more costly than your funding will allow. There might also be some response bias: Some students might be hesitant about criticizing the faculty (while others might be far too eager to do so).

3.131. Use a block design: Separate men and women, and randomly allocate each gender among the six treatments.

The remaining exercises relate to the material of Section 3.4 (Ethics). Answers are given for the first two; the rest call for student opinions, or information specific to the student's institution.

3.133. The latter method (CASI) will show a higher percentage of drug use because respondents will generally be more comfortable (and more assured of anonymity) about revealing embarrassing or illegal behavior to a computer than to a person, so they will be more likely to be honest.

Chapter 4 Solutions

4.1. Only 6 of the first 20 digits on line 119 correspond to "heads," so the proportion of heads is $\frac{6}{20} = 0.3$. With such a small sample, random variation can produce results different from the expected value (0.5).

```
95857   07118   87664   92099
TTTTT   HTHHT   TTTTH   THHTT
```

4.3. (a) Most answers (99.5% of them) will be between 82% and 98%. **(b)** Based on 100,000 simulated trials—more than students are expected to do—the longest string of misses will be quite short (3 or fewer with probability 99%, 5 or fewer with probability 99.99%). The average ("expected") longest run of misses is about 1.7. For shots made, the average run is about 27, but there is lots of variation; 77% of simulations will have a longest run of made shots between 17 and 37, and about 95% of simulation will fall between 12 and 46.

4.5. If you hear music (or talking) one time, you will almost certainly hear the same thing for several more checks after that. (For example, if you tune in at the beginning of a 5-minute song and check back every 5 seconds, you'll hear that same song over 30 times.)

4.7. Out of a very large number of patients taking this medication, the fraction who experience this bad side effect is about 0.00001.

    Note: *Student explanations will vary, but should make clear that 0.00001 is a long-run average rate of occurrence. Because a probability of 0.00001 is often stated as "1 in in 100,000," it is tempting to interpret this probability as meaning "exactly 1 out of every 100,000." While we* expect *about 1 occurrence of side effects out of 100,000 patients, the actual number of side effects patients is random; it might be 0, or 1, or 2, ...*

4.9. The true probability (assuming perfectly fair dice) is $1 - \left(\frac{5}{6}\right)^4 \doteq 0.5177$, so students should conclude that the probability is "quite close to 0.5."

4.11. One possibility: from 0 to __ hours (the largest number should be big enough to include all possible responses). In addition, some students might respond with fractional answers (e.g., 3.5 hours).

4.13. $P(\text{Blue, Green, Black, Brown, Grey, or White}) = 1 - P(\text{Purple, Red, Orange, or Yellow}) = 1 - (0.14 + 0.08 + 0.05 + 0.03) = 1 - 0.3 = 0.7$. Using Rule 4 (the complement rule) is slightly easier, because we only need to add the four probabilities of the colors we do *not* want, rather than adding the six probabilities of the colors we want.

4.15. In Example 4.13, $P(B) = P(6 \text{ or greater})$ was found to be 0.222, so $P(A \text{ or } B) = P(A) + P(B) = 0.301 + 0.222 = 0.523$.

4.17. If T_k is the event "get tails on the kth flip," then T_1 and T_2 are independent, and $P(\text{two tails}) = P(T_1 \text{ and } T_2) = P(T_1)P(T_2) = \left(\frac{1}{2}\right)\left(\frac{1}{2}\right) = \frac{1}{4}$.

4.19. (a) The probability that both of two disjoint events occur is 0. (Multiplication is appropriate for *independent* events.) **(b)** Probabilities must be no more than 1; $P(A \text{ and } B)$ will be no more than 0.5. (We cannot determine this probability exactly from the given information.) **(c)** $P(A^c) = 1 - 0.35 = 0.65$.

4.21. There are six possible outcomes: { link1, link2, link3, link4, link5, leave }.

4.23. (a) $P(\text{"Empire State of Mind" or "I Gotta Feeling"}) = 0.180 + 0.068 = 0.248$.
(b) $P(\text{neither "Empire State of Mind" nor "I Gotta Feeling"}) = 1 - 0.248 = 0.752$.

4.25. (a) The given probabilities have sum 0.97, so $P(\text{type AB}) = 0.03$.
(b) $P(\text{type O or B}) = 0.44 + 0.11 = 0.55$.

4.27. (a) Not legitimate because the probabilities sum to 2. **(b)** Legitimate (for a nonstandard deck). **(c)** Legitimate (for a nonstandard die).

4.29. (a) The given probabilities have sum 0.72, so this probability must be 0.28.
(b) $P(\text{at least a high school education}) = 1 - P(\text{has not finished HS}) = 1 - 0.12 = 0.88$. (Or add the other three probabilities.)

4.31. For example, the probability for A-positive blood is $(0.42)(0.84) = 0.3528$ and for A-negative $(0.42)(0.16) = 0.0672$.

Blood type	A+	A−	B+	B−	AB+	AB−	O+	O−
Probability	0.3528	0.0672	0.0924	0.0176	0.0252	0.0048	0.3696	0.0704

4.33. (a) There are six arrangements of the digits 4, 9, and 1 (491, 419, 941, 914, 149, 194), so that $P(\text{win}) = \frac{6}{1000} = 0.006$. **(b)** The only winning arrangement is 222, so $P(\text{win}) = \frac{1}{1000} = 0.001$.

4.35. $P(\text{none are O-negative}) = (1 - 0.07)^{10} \doteq 0.4840$, so $P(\text{at least one is O-negative}) \doteq 1 - 0.4840 = 0.5160$.

4.37. This computation would only be correct if the events "a randomly selected person is at least 75" and "a randomly selected person is a woman" were independent. This is likely not true; in particular, as women have a greater life expectancy than men, this fraction is probably greater than 3%.

4.39. (a) $(0.65)^3 \doteq 0.2746$ (under the random walk theory). **(b)** 0.35 (because performance in separate years is independent). **(c)** $(0.65)^2 + (0.35)^2 = 0.545$.

4.41. Note that $A = (A \text{ and } B) \text{ or } (A \text{ and } B^c)$, and the events $(A \text{ and } B)$ and $(A \text{ and } B^c)$ are disjoint, so Rule 3 says that

$$P(A) = P\big((A \text{ and } B) \text{ or } (A \text{ and } B^c)\big) = P(A \text{ and } B) + P(A \text{ and } B^c).$$

If $P(A \text{ and } B) = P(A)P(B)$, then we have $P(A \text{ and } B^c) = P(A) - P(A)P(B) = P(A)(1 - P(B))$, which equals $P(A)P(B^c)$ by the complement rule.

4.43. (a) Nancy and David's children can have alleles BB, BO, or OO, so they can have blood type B or O. (The table on the right shows the possible combinations.) **(b)** Either note that the four combinations in the table are equally likely or compute $P(\text{type O}) = P(\text{O from Nancy and O from David}) = 0.5^2 = 0.25$ and $P(\text{type B}) = 1 - P(\text{type O}) = 0.75$.

	B	O
B	BB	BO
O	BO	OO

4.45. (a) Any child of Jasmine and Joshua has an equal (1/4) chance of having blood type AB, A, B, or O (see the allele combinations in the table). Therefore, $P(\text{type O}) = 0.25$. **(b)** $P(\text{all three have type O}) = 0.25^3 = 0.015625 = \frac{1}{64}$. $P(\text{first has type O, next two do not}) = 0.25 \cdot 0.75^2 = 0.140625 = \frac{9}{64}$.

	A	O
B	AB	BO
O	AO	OO

4.47. If H is the number of heads, then the distribution of H is as given on the right. $P(H = 0)$, the probability of two tails was previously computed in Exercise 4.17.

Value of H	0	1	2
Probabilities	1/4	1/2	1/4

4.49. (a) The probabilities for a discrete *random variable* always add to one. **(b)** Continuous random variables can take values from any interval, not just 0 to 1. **(c)** A Normal random variable is continuous. (Also, a distribution is *associated with* a random variable, but "distribution" and "random variable" are not the same things.)

4.51. (a) Based on the information from Exercise 4.50, along with the complement rule, $P(T) = 0.19$ and $P(T^c) = 0.81$. **(b)** Use the multiplication rule for independent events; for example, $P(TTT) = 0.19^3 \doteq 0.0069$, $P(TTT^c) = (0.19^2)(0.81) \doteq 0.0292$, $P(TT^cT^c) = (0.19)(0.81^2) \doteq 0.1247$, and $P(T^cT^cT^c) = 0.81^3 \doteq 0.5314$. **(c)** Add up the probabilities from (b) that correspond to each value of X.

Outcome	TTT	TTT^c	TT^cT	T^cTT	T^cT^cT	T^cTT^c	TT^cT^c	$T^cT^cT^c$
Probability	0.0069	0.0292	0.0292	0.0292	0.1247	0.1247	0.1247	0.5314
Value of X	0	1			2			3
Probability	0.0069	0.0877			0.3740			0.5314

4.53. (a) See also Exercise 4.22. If we view this time as being measured to any degree of accuracy, it is continuous; if it is rounded, it is discrete. **(b)** A count like this must be a whole number, so it is discrete. **(c)** Incomes—whether given in dollars and cents, or rounded to the nearest dollar—are discrete. (However, it is often useful to treat such variables as continuous.)

4.55. (a) Histogram on the right. **(b)** "At least one nonword error" is the event "$X \geq 1$" (or "$X > 0$"). $P(X \geq 1) =$ $1 - P(X = 0) = 0.9$. **(c)** "$X \leq 2$" is "no more than two nonword errors," or "fewer than three nonword errors."

$$P(X \leq 2) = 0.7 = P(X = 0) + P(X = 1) + P(X = 2)$$
$$= 0.1 + 0.3 + 0.3$$
$$P(X < 2) = 0.4 = P(X = 0) + P(X = 1) = 0.1 + 0.3$$

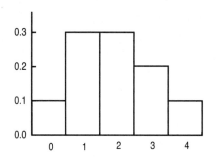

4.57. (a) The pairs are given below. We must assume that we can distinguish between, for example, "(1,2)" and "(2,1)"; otherwise, the outcomes are not equally likely. **(b)** Each pair has probability 1/36. **(c)** The value of X is given below each pair. For the distribution (given below), we see (for example) that there are four pairs that add to 5, so $P(X = 5) = \frac{4}{36}$. Histogram below, right. **(d)** $P(7 \text{ or } 11) = \frac{6}{36} + \frac{2}{36} = \frac{8}{36} = \frac{2}{9}$. **(e)** $P(\text{not } 7) = 1 - \frac{6}{36} = \frac{5}{6}$.

(1,1)	(1,2)	(1,3)	(1,4)	(1,5)	(1,6)
2	3	4	5	6	7
(2,1)	(2,2)	(2,3)	(2,4)	(2,5)	(2,6)
3	4	5	6	7	8
(3,1)	(3,2)	(3,3)	(3,4)	(3,5)	(3,6)
4	5	6	7	8	9
(4,1)	(4,2)	(4,3)	(4,4)	(4,5)	(4,6)
5	6	7	8	9	10
(5,1)	(5,2)	(5,3)	(5,4)	(5,5)	(5,6)
6	7	8	9	10	11
(6,1)	(6,2)	(6,3)	(6,4)	(6,5)	(6,6)
7	8	9	10	11	12

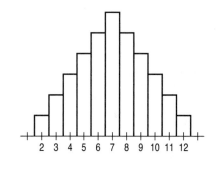

Value of X	2	3	4	5	6	7	8	9	10	11	12
Probability	$\frac{1}{36}$	$\frac{2}{36}$	$\frac{3}{36}$	$\frac{4}{36}$	$\frac{5}{36}$	$\frac{6}{36}$	$\frac{5}{36}$	$\frac{4}{36}$	$\frac{3}{36}$	$\frac{2}{36}$	$\frac{1}{36}$

4.59. The table on the right shows the additional columns to add to the table given in the solution to Exercise 4.57. There are 48 possible (equally-likely) combinations.

Value of X	2	3	4	5	6	7	8	9	10	11	12	13	14
Probability	$\frac{1}{48}$	$\frac{2}{48}$	$\frac{3}{48}$	$\frac{4}{48}$	$\frac{5}{48}$	$\frac{6}{48}$	$\frac{6}{48}$	$\frac{6}{48}$	$\frac{5}{48}$	$\frac{4}{48}$	$\frac{3}{48}$	$\frac{2}{48}$	$\frac{1}{48}$

(1,7)	(1,8)
8	9
(2,7)	(2,8)
9	10
(3,7)	(3,8)
10	11
(4,7)	(4,8)
11	12
(5,7)	(5,8)
12	13
(6,7)	(6,8)
13	14

4.61. (a) $P(X < 0.6) = 0.6$. **(b)** $P(X \leq 0.6) = 0.6$. **(c)** For continuous random variables, "equal to" has no effect on the probability; that is, $P(X = c) = 0$ for any value of c.

4.63. (a) The height should be $\frac{1}{2}$ since the area under the curve must be 1. The density curve is at the right. **(b)** $P(Y \le 1.6) = \frac{1.6}{2} = 0.8$. **(c)** $P(0.5 < Y < 1.7) = \frac{1.2}{2} = 0.6$. **(d)** $P(Y \ge 0.95) = \frac{1.05}{2} = 0.525$.

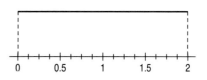

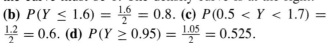

4.65. $P(8 \le \bar{x} \le 10) = P\left(\frac{8-9}{0.0724} \le \frac{\bar{x}-9}{0.0724} \le \frac{10-9}{0.0724}\right) = P(-13.8 \le Z \le 13.8)$. This probability is essentially 1; $\bar{x}$ will almost certainly estimate μ within ± 1 (in fact, it will almost certainly be much closer than this).

4.67. The possible values of X are \$0 and \$1, each with probability 0.5 (because the coin is fair). The mean is $\$0 \left(\frac{1}{2}\right) + \$1 \left(\frac{1}{2}\right) = \$0.50$.

4.69. If $Y = 15 + 8X$, then $\mu_Y = 15 + 8\mu_X = 15 + 8(10) = 95$.

4.71. First we note that $\mu_X = 0(0.5) + 2(0.5) = 1$, so $\sigma_X^2 = (0-1)^2(0.5) + (2-1)^2(0.5) = 1$ and $\sigma_X = \sqrt{\sigma_X^2} = 1$.

4.73. The mean is
$$\mu_X = (0)(0.3) + (1)(0.1) + (2)(0.1) + (3)(0.2) + (4)(0.1) + (5)(0.2) = 2.3 \text{ servings.}$$
The variance is
$$\sigma_X^2 = (0 - 2.3)^2(0.3) + (1 - 2.3)^2(0.1) + (2 - 2.3)^2(0.1)$$
$$+ (3 - 2.3)^2(0.2) + (4 - 2.3)^2(0.1) + (5 - 2.3)^2(0.2) = 3.61,$$
so the standard deviation is $\sigma_X = \sqrt{3.61} = 1.9$ servings.

4.75. The average grade is $\mu = (0)(0.05) + (1)(0.04) + (2)(0.20) + (3)(0.40) + (4)(0.31) = 2.88$.

4.77. In the solution to Exercise 4.74, we found $\mu_X = 0.1538$ aces, so
$$\sigma_X^2 = (0 - 0.1538)^2(0.8507) + (1 - 0.1538)^2(0.1448) + (2 - 0.1538)^2(0.0045) \doteq 0.1391,$$
and the standard deviation is $\sigma_X \doteq \sqrt{0.1391} \doteq 0.3730$ aces.

4.79. (a) With $\rho = 0$, the variance is $\sigma_{X+Y}^2 = \sigma_X^2 + \sigma_Y^2 = (75)^2 + (41)^2 = 7306$, so the standard deviation is $\sigma_{X+Y} = \sqrt{7306} \doteq \$85.48$. **(b)** This is larger; the negative correlation decreased the variance.

4.81. The situation described in this exercise—"people who have high intakes of calcium in their diets are more compliant than those who have low intakes"—implies a positive correlation between calcium intake and compliance. Because of this, the variance of total calcium intake is greater than the variance we would see if there were no correlation (as the calculations in Example 4.38 demonstrate).

4.83. (a) The mean for one coin is $\mu_1 = (0)\left(\frac{1}{2}\right) + (1)\left(\frac{1}{2}\right) = 0.5$ and the variance is $\sigma_1^2 = (0 - 0.5)^2 \left(\frac{1}{2}\right) + (1 - 0.5)^2 \left(\frac{1}{2}\right) = 0.25$, so the standard deviation is $\sigma_1 = 0.5$.

(b) Multiply μ_1 and σ_1^2 by 4: $\mu_4 = 4\mu_1 = 2$ and $\sigma_4^2 = 4\sigma_1^4 = 1$, so $\sigma_4 = 1$. **(c)** Note that because of the symmetry of the distribution, we do not need to compute the mean to see that $\mu_4 = 2$; this is the obvious balance point of the probability histogram in Figure 4.7. The details of the two computations are

$$\mu_W = (0)(0.0625) + (1)(0.25) + (2)(0.375) + (3)(0.25) + (4)(0.0625) = 2$$
$$\sigma_W^2 = (0-2)^2(0.0625) + (1-2)^2(0.25)$$
$$+ (2-2)^2(0.375) + (3-2)^2(0.25) + (4-2)^2(0.0625) = 1.$$

4.85. With R as the rod length and B_1 and B_2 the bearing lengths, we have $\mu_{B_1+R+B_2} = 12 + 2 \cdot 2 = 16$ cm and $\sigma_{B_1+R+B_2} = \sqrt{0.004^2 + 2 \cdot 0.001^2} \doteq 0.004243$ mm.

4.87. (a) Not independent: Knowing the total X of the first two cards tells us something about the total Y for three cards. **(b)** Independent: Separate rolls of the dice should be independent.

4.89. With $\rho = 1$, we have:

$$\sigma_{X+Y}^2 = \sigma_X^2 + \sigma_Y^2 + 2\rho\sigma_X\sigma_Y = \sigma_X^2 + \sigma_Y^2 + 2\sigma_X\sigma_Y = (\sigma_X + \sigma_Y)^2$$

And of course, $\sigma_{X+Y} = \sqrt{(\sigma_X + \sigma_Y)^2} = \sigma_X + \sigma_Y$.

4.91. Although the probability of having to pay for a total loss for one or more of the 10 policies is very small, if this were to happen, it would be financially disastrous. On the other hand, for thousands of policies, the law of large numbers says that the average claim on many policies will be close to the mean, so the insurance company can be assured that the premiums they collect will (almost certainly) cover the claims.

4.93. (a) Add up the given probabilities and subtract from 1; this gives P(man does not die in the next five years) $= 0.99749$. **(b)** The distribution of income (or loss) is given below. Multiplying each possible value by its probability gives the mean intake $\mu \doteq \$623.22$.

Age at death	21	22	23	24	25	Survives
Loss or income	−\$99,825	−\$99,650	−\$99,475	−\$99,300	−\$99,125	\$875
Probability	0.00039	0.00044	0.00051	0.00057	0.00060	0.99749

4.95. The events "roll a 3" and "roll a 5" are disjoint, so $P(3 \text{ or } 5) = P(3) + P(5) = \frac{1}{6} + \frac{1}{6} = \frac{2}{6} = \frac{1}{3}$.

4.97. Let A be the event "next card is an ace" and B be "two of Slim's four cards are aces." Then, $P(A \mid B) = \frac{2}{48}$ because (other than those in Slim's hand) there are 48 cards, of which 2 are aces.

4.99. This computation uses the addition rule for disjoint events, which is appropriate for this setting because B (full-time students) is made up of four disjoint groups (those in each of the four age groups).

4.101. The tree diagram shows the probability found in Exercise 4.98 on the top branch. The middle two branches (added together) give the probability that Slim gets exactly one diamond from the next two cards, and the bottom branch is the probability that neither card is a diamond.

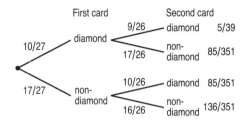

4.103. For a randomly chosen adult, let S = "(s)he gets enough sleep" and let E = "(s)he gets enough exercise," so $P(S) = 0.4$, $P(E) = 0.46$, and $P(S \text{ and } E) = 0.24$.
(a) $P(S \text{ and } E^c) = 0.4 - 0.24 = 0.16$. **(b)** $P(S^c \text{ and } E) = 0.46 - 0.24 = 0.22$.
(c) $P(S^c \text{ and } E^c) = 1 - (0.4 + 0.46 - 0.24) = 0.38$. **(d)** The answers in (a) and (b) are found by a variation of the addition rule for disjoint events: We note that $P(S) = P(S \text{ and } E) + P(S \text{ and } E^c)$ and $P(E) = P(S \text{ and } E) + P(S^c \text{ and } E)$. In each case, we know the first two probabilities, and find the third by subtraction. The answer for (c) is found by using the general addition rule to find $P(S \text{ or } E)$, and noting that $S^c \text{ and } E^c = (S \text{ or } E)^c$.

4.105. For a randomly chosen high school student, let L = "student admits to lying" and M = "student is male," so $P(L) = 0.48$, $P(M) = 0.5$, and $P(M \text{ and } L) = 0.25$. Then $P(M \text{ or } L) = P(M) + P(L) - P(M \text{ and } L) = 0.73$.

4.107. Let B = "student is a binge drinker" and M = "student is male." **(a)** The four probabilities sum to $0.11 + 0.12 + 0.32 + 0.45 = 1$. **(b)** $P(B^c) = 0.32 + 0.45 = 0.77$.
(c) $P(B^c \mid M) = \frac{P(B^c \text{ and } M)}{P(M)} = \frac{0.32}{0.11 + 0.32} \doteq 0.7442$. **(d)** In the language of this chapter, the events are not independent. An attempt to phrase this for someone who has not studied this material might say something like, "Knowing a student's gender gives some information about whether or not that student is a binge drinker."

Note: *Specifically, male students are slightly more likely to be binge drinkers. This statement might surprise students who look at the table and note that the proportion of binge drinkers in the men's columns is smaller than that proportion in the women's column. We cannot compare those proportions directly; we need to compare the conditional probabilities of binge drinkers within each given gender (see the solution to the next exercise.)*

4.109. Let M = "male" and C = "attends a 4-year institution." (C is not an obvious choice, but it is less confusing than F, which we might mistake for "female.") We have been given

	Men	Women
4-year	0.2684	0.3416
2-year	0.1599	0.2301

$P(C) = 0.61$, $P(C^c) = 0.39$, $P(M \mid C) = 0.44$ and $P(M \mid C^c) = 0.41$. **(a)** To create the table, observe that:

$$P(M \text{ and } C) = P(M \mid C)P(C) = (0.44)(0.61) = 0.2684$$

And similarly, $P(M \text{ and } C^c) = P(M \mid C^c)P(C^c) = (0.41)(0.39) = 0.1599$. The other two entries can be found in a similar fashion or by observing that, for example, the two numbers on the first row must sum to $P(C) = 0.61$.
(b) $P(C \mid M^c) = \frac{P(C \text{ and } M^c)}{P(M^c)} = \frac{0.3416}{0.3416 + 0.2301} \doteq 0.5975$.

4.111. As before, let M = "male" and C = "attends a 4-year institution." For this tree diagram, we need to compute $P(M) = 0.2684 + 0.1599 = 0.4283$, $P(M^c) = 0.3416 + 0.2301 = 0.5717$, as well as $P(C \mid M)$, $P(C \mid M^c)$, $P(C^c \mid M)$, and $P(C^c \mid M^c)$. For example,

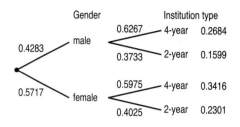

$$P(C \mid M) = \frac{P(C \text{ and } M)}{P(M)} = \frac{0.2684}{0.4283} \doteq 0.6267.$$

All the computations for this diagram are "inconvenient" because they require that we work *backward* from the ending probabilities, instead of working *forward* from the given probabilities (as we did in the previous tree diagram).

4.113. $P(A \mid B) = \frac{P(A \text{ and } B)}{P(B)} = \frac{0.082}{0.261} \doteq 0.3142$. If A and B were independent, then $P(A \mid B)$ would equal $P(A)$, and also $P(A \text{ and } B)$ would equal the product $P(A)P(B)$.

4.115. (a) "The vehicle is a light truck" = A^c; $P(A^c) = 0.69$. **(b)** "The vehicle is an imported car" = A and B. To find this

	$P(A) = 0.31$	$P(A^c) = \mathbf{0.69}$
$P(B) = 0.22$	$P(A \text{ and } B) = 0.08$	$P(A^c \text{ and } B) = 0.14$
$P(B^c) = \mathbf{0.78}$	$P(A \text{ and } B^c) = 0.23$	$P(A^c \text{ and } B^c) = \mathbf{0.55}$

probability, note that we have been given $P(B^c) = 0.78$ and $P(A^c \text{ and } B^c) = 0.55$. From this we can determine that $78\% - 55\% = 23\%$ of vehicles sold were domestic cars—that is, $P(A \text{ and } B^c) = 0.23$—so $P(A \text{ and } B) = P(A) - P(A \text{ and } B^c) = 0.31 - 0.23 = 0.08$.

   Note: *The table shown here summarizes all that we can determine from the given information* (**bold**).

4.117. See also the solution to Exercise 4.115, especially the table of probabilities given there. **(a)** $P(A^c \mid B) = \frac{P(A^c \text{ and } B)}{P(B)} = \frac{0.14}{0.22} \doteq 0.6364$. **(b)** The events A^c and B are *not* independent; if they were, $P(A^c \mid B)$ would be the same as $P(A^c) = 0.69$.

4.119. We seek $P(\text{at least one offer}) = P(A \text{ or } B \text{ or } C)$; we can find this as $1 - P(\text{no offers}) = 1 - P(A^c \text{ and } B^c \text{ and } C^c)$. We see in the Venn diagram of Exercise 4.118 that this probability is 1.

4.121. $P(B \mid C) = \frac{P(B \text{ and } C)}{P(C)} = \frac{0.1}{0.3} = \frac{1}{3}$. $P(C \mid B) = \frac{P(B \text{ and } C)}{P(B)} = \frac{0.1}{0.5} = 0.2$.

4.123. Let M be the event "the person is a man" and B be "the person earned a bachelor's degree." **(a)** $P(M) = \frac{991}{2403} \doteq 0.4124$. Or take the answer from part (b) of the previous exercise and subtract from 1. **(b)** $P(B \mid M) = \frac{661/2403}{991/2403} = \frac{661}{991} \doteq 0.6670$. (This is the "Bachelor's" entry from the "Male" row, divided by that row's total.) **(c)** $P(M \text{ and } B) = P(M) P(B \mid M) \doteq (0.4124)(0.6670) \doteq 0.2751$. This agrees with the directly computed probability: $P(M \text{ and } B) = \frac{661}{2403} \doteq 0.2751$.

4.125. **(a)** Add up the numbers in the first and second columns. We find that there are 186,210 thousand (i.e., over 186 million) people aged 25 or older, of which 125,175 thousand are in the labor force, so $P(L) = \frac{125,175}{186,210} \doteq 0.6722$. **(b)** $P(L \mid C) = \frac{P(L \text{ and } C)}{P(C)} = \frac{40,414}{51,568} \doteq 0.7837$. **(c)** L and C are *not* independent; if they were, the two probabilities in (a) and (b) would be equal.

4.127. The population includes retired people who have left the labor force. Retired persons are more likely than other adults to have not completed high school; consequently, a relatively large number of retired persons fall in the "did not finish high school" category.

   Note: *Details of this lurking variable can be found in the Current Population Survey annual report on "Educational Attainment in the United States." For 2006, this report says that among the 65-and-over population, about 24.8% have not completed high school, compared to about 19.3% of the under-65 group.*

4.129. **(a)** Her brother has type aa, and he got one allele from each parent. But neither parent is albino, so neither could be type aa. **(b)** The table on the right shows the possible combinations, each of which is equally likely, so $P(aa) = 0.25$, $P(Aa) = 0.5$, and $P(AA) = 0.25$. **(c)** Beth is either AA or Aa, and $P(AA \mid \text{not } aa) = \frac{0.25}{0.75} = \frac{1}{3}$, while $P(Aa \mid \text{not } aa) = \frac{0.50}{0.75} = \frac{2}{3}$.

	A	a
A	AA	Aa
a	Aa	aa

4.131. Let C be the event "Toni is a carrier," T be the event "Toni tests positive," and D be "her son has DMD." We have $P(C) = \frac{2}{3}$, $P(T \mid C) = 0.7$, and $P(T \mid C^c) = 0.1$. Therefore, $P(T) = P(T \text{ and } C) + P(T \text{ and } C^c) = P(C)\,P(T \mid C) + P(C^c)\,P(T \mid C^c) = \left(\frac{2}{3}\right)(0.7) + \left(\frac{1}{3}\right)(0.1) = 0.5$, and:

$$P(C \mid T) = \frac{P(T \text{ and } C)}{P(C)} = \frac{(2/3)(0.7)}{0.5} = \frac{14}{15} \doteq 0.9333$$

4.133. The mean of X is $\mu_X = (-1)(0.3) + (2)(0.7) = 1.1$, so the mean if many such values will be close to 1.1.

4.135. **(a)** Because the possible values of X are 1 and 2, the possible values of Y are $3 \cdot 1^2 - 2 = 1$ (with probability 0.4) and $3 \cdot 2^2 - 2 = 10$ (with probability 0.6). **(b)** $\mu_Y = (1)(0.4) + (10)(0.6) = 6.4$ and $\sigma_Y^2 = (1 - 6.4)^2(0.4) + (10 - 6.4)^2(0.6) = 19.44$, so $\sigma_X = \sqrt{19.44} \doteq 4.4091$. **(c)** Those rule are for transformations of the form $aX + b$, not $aX^2 + b$.

4.137. **(a)** $P(A) = \frac{5}{36}$ and $P(B) = \frac{10}{36} = \frac{5}{18}$. **(b)** $P(A) = \frac{5}{36}$ and $P(B) = \frac{21}{36} = \frac{7}{12}$. **(c)** $P(A) = \frac{15}{36} = \frac{5}{12}$ and $P(B) = \frac{10}{36} = \frac{5}{18}$. **(d)** $P(A) = \frac{15}{36} = \frac{5}{12}$ and $P(B) = \frac{10}{36} = \frac{5}{18}$.

4.139. For each bet, the mean is the winning probability times the winning payout, plus the losing probability times $-\$10$. These are summarized in the table on the right; all mean payoffs equal $\$0$.

Point	Expected Payoff
4 or 10	$\frac{1}{3}(+\$20) + \frac{2}{3}(-\$10) = 0$
5 or 9	$\frac{2}{5}(+\$15) + \frac{3}{5}(-\$10) = 0$
6 or 8	$\frac{5}{11}(+\$12) + \frac{6}{11}(-\$10) = 0$

Note: *Alternatively, we can find the mean amount of money we have at the end of the bet. For example, if the point is 4 or 10, we end with either $30 or $0, and our expected ending amount is $\frac{1}{3}(\$30) + \frac{2}{3}(\$0) = \$10$—equal to the amount of the bet.*

4.141. (a) All probabilities are greater than or equal to 0, and their sum is 1. **(b)** Let R_1 be Taster 1's rating and R_2 be Taster 2's rating. Add the probabilities on the diagonal (upper left to lower right): $P(R_1 = R_2) = 0.03 + 0.07 + 0.25 + 0.20 + 0.06 = 0.61$. **(c)** $P(R_1 > 3) = 0.39$ (the sum of the ten numbers in the bottom two rows), and $P(R_2 > 3) = 0.39$ (the sum of the ten numbers in the right two columns).

4.143. This is the probability of 19 (independent) losses, followed by a win; by the multiplication rule, this is $0.994^{19} \cdot 0.006 \doteq 0.005352$.

4.145. With B, M, and D representing the three kinds of degrees, and W meaning the degree recipient was a woman, we have been given:

$$P(B) = 0.71, \qquad P(M) = 0.23, \qquad P(D) = 0.06,$$
$$P(W \mid B) = 0.44, \quad P(W \mid M) = 0.41, \quad P(W \mid D) = 0.30.$$

Therefore, we find

$$P(W) = P(W \text{ and } B) + P(W \text{ and } M) + P(W \text{ and } D)$$
$$= P(B)\,P(W \mid B) + P(M)\,P(W \mid M) + P(D)\,P(W \mid D) = 0.4247,$$

so:

$$P(B \mid W) = \frac{P(B \text{ and } W)}{P(W)} = \frac{P(B)\,P(W \mid B)}{P(W)} = \frac{0.3124}{0.4247} \doteq 0.7356$$

4.147. P(no point is established) $= \frac{12}{36} = \frac{1}{3}$.

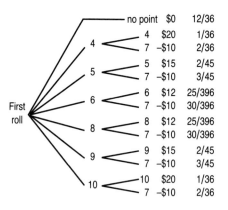

In Exercise 4.139, the probabilities of winning each odds bet were given as $\frac{1}{3}$ for 4 and 10, $\frac{2}{5}$ for 5 and 9, and $\frac{5}{11}$ for 6 and 8. This tree diagram can get a bit large (and crowded). In the diagram shown on the right, the probabilities are omitted from the individual branches. The probability of winning an odds bet on 4 or 10 (with a net payout of $20) is $\left(\frac{3}{36}\right)\left(\frac{1}{3}\right) = \frac{1}{36}$. Losing that odds bet costs $10, and has probability $\left(\frac{3}{36}\right)\left(\frac{2}{3}\right) = \frac{2}{36}$ (or $\frac{1}{18}$). Similarly, the probability of winning an odds bet on 5 or 9 is $\left(\frac{4}{36}\right)\left(\frac{2}{5}\right) = \frac{2}{45}$, and the probability of losing that bet is $\left(\frac{4}{36}\right)\left(\frac{3}{5}\right) = \frac{3}{45}$ (or $\frac{1}{15}$). For an odds bet on 6 or 8, we win $12 with probability $\left(\frac{5}{36}\right)\left(\frac{5}{11}\right) = \frac{25}{396}$, and lose $10 with probability $\left(\frac{5}{36}\right)\left(\frac{6}{11}\right) = \frac{30}{396}$ (or $\frac{5}{66}$).

To confirm that this game is fair, one can multiply each payoff by its probability then add up all of those products. More directly, because each individual odds bet is fair (as was shown in the solution to Exercise 4.139), one can argue that taking the odds bet whenever it is available must be fair.

4.149. Let R_1 be Taster 1's rating and R_2 be Taster 2's rating. $P(R_1 = 3) = 0.01 + 0.05 + 0.25 + 0.05 + 0.01 = 0.37$, so:

$$P(R_2 > 3 \mid R_1 = 3) = \frac{P(R_2 > 3 \text{ and } R_1 = 3)}{P(R_1 = 3)} = \frac{0.05 + 0.01}{0.37} \doteq 0.1622$$

4.151. The event $\{Y < 1/2\}$ is the bottom half of the square, while $\{Y > X\}$ is the upper left triangle of the square. They overlap in a triangle with area $1/8$, so:

$$P(Y < \tfrac{1}{2} \mid Y > X) = \frac{P(Y < \tfrac{1}{2} \text{ and } Y > X)}{P(Y > X)} = \frac{1/8}{1/2} = \frac{1}{4}$$

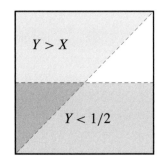

Chapter 5 Solutions

5.1. The population is iPhone users (or iPhone users who use the AppsFire service). The statistic is an average of 65 apps per device. Likely values will vary, in part based on how many apps are on student phones (which they might consider "typical").

5.3. When $n = 144$, the mean is $\mu_{\bar{x}} = \mu = 240$ (unchanged), and the standard deviation is $\sigma_{\bar{x}} = \sigma/\sqrt{n} = 1.5$. Increasing n does not change $\mu_{\bar{x}}$ but decreases $\sigma_{\bar{x}}$, the variability of the sampling distribution. (In this case, because n was increased by a factor of 4, $\sigma_{\bar{x}}$ was halved.)

5.5. When $n = 1296$, $\sigma_{\bar{x}} = \sigma/\sqrt{1296} = 18/36 = 0.5$. The sampling distribution of $\bar{x}$ is approximately $N(240, 0.5)$, so about 95% of the time, $\bar{x}$ is between 239 and 241.

5.7. **(a)** Either change "variance" to "standard deviation" (twice), or change the formula at the end to $10^2/30$. **(b)** Standard deviation decreases with increasing sample size. **(c)** $\mu_{\bar{x}}$ always equals μ, regardless of the sample size.

5.9. **(a)** $\mu = 3388/10 = 338.8$. **(b)** The scores will vary depending on the starting row. The smallest and largest possible means are 290 and 370. **(c)** Answers will vary. Shown on the right is a histogram of the (exact) sampling distribution. With a sample size of only 3, the distribution is noticeably non-Normal. **(d)** The center of the exact sampling distribution is μ, but with only 10 values of $\bar{x}$, this might not be true for student histograms.

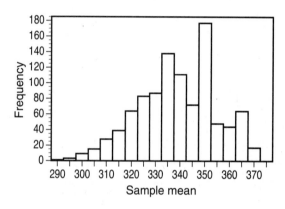

Note: *This histograms were found by considering all 1000 possible samples.*

5.11. **(a)** With $n = 200$, the 95% probability range was about ± 10 minutes, so need a larger sample size. (Specifically, to halve the range, we need to roughly quadruple the sample size.) **(b)** We need $2\sigma_{\bar{x}} = \frac{5}{60}$, so $\sigma_{\bar{x}} \doteq 0.04167$. **(c)** With $\sigma = 1.15$, we have $\sqrt{n} = \frac{1.15}{0.04167} = 27.6$, so $n = 761.76$—use 762 students.

5.13. Mean $\mu = 250$ ml and standard deviation $\sigma/\sqrt{6} = 3/\sqrt{6} \doteq 1.2247$ ml.

5.15. In Exercise 5.13, we found that $\sigma_{\bar{x}} \doteq 1.2247$ ml, so $\bar{x}$ has a $N(250$ ml, 1.2247 ml$)$ distribution. **(a)** On the right. The Normal curve for $\bar{x}$ should be taller by a factor of $\sqrt{6}$ and skinnier by a factor of $1/\sqrt{6}$. **(b)** The probability that a single can's volume differs from the target by at least 1 ml—one-third of a standard deviation—is $1 - P(-0.33 < Z < 0.33) = 0.7414$. **(c)** The probability that $\bar{x}$ is at least 1 ml from the target is

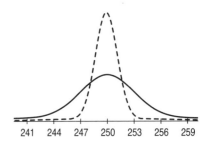

$$1 - P(249 < \bar{x} < 251) = 1 - P(-0.82 < Z < 0.82) = 0.4122.$$

5.17. (a) $\bar{x}$ is not systematically higher than or lower than μ; that is, it has no particular tendency to underestimate or overestimate μ. (In other words, it is "just right" on the average.) **(b)** With large samples, $\bar{x}$ is more likely to be close to μ because with a larger sample comes more information (and therefore less uncertainty).

5.19. (a) $\mu_{\bar{x}} = 0.5$ and $\sigma_{\bar{x}} = \sigma/\sqrt{50} = 0.7/\sqrt{50} \doteq 0.09899$. **(b)** Because this distribution is only approximately Normal, it would be quite reasonable to use the 68–95–99.7 rule to give a rough estimate: 0.6 is about one standard deviation above the mean, so the probability should be about 0.16 (half of the 32% that falls outside ± 1 standard deviation). Alternatively, $P(\bar{x} > 0.6) \doteq P\left(Z > \frac{0.6 - 0.5}{0.09899}\right) = P(Z > 1.01) = 0.1562$.

5.21. Let X be Sheila's measured glucose level. **(a)** $P(X > 140) = P(Z > 1.5) = 0.0668$. **(b)** If $\bar{x}$ is the mean of three measurements (assumed to be independent), then $\bar{x}$ has a $N(125, 10/\sqrt{3})$ or $N(125$ mg/dl, 5.7735 mg/dl$)$ distribution, and $P(\bar{x} > 140) = P(Z > 2.60) = 0.0047$.

5.23. The mean of three measurements has a $N(125$ mg/dl, 5.7735 mg/dl$)$ distribution, and $P(Z > 1.645) = 0.05$ if Z is $N(0, 1)$, so $L = 125 + 1.645 \cdot 5.7735 \doteq 134.5$ mg/dl.

5.25. If W is total weight, and $\bar{x} = W/25$, then:

$$P(W > 5200) = P(\bar{x} > 208) \doteq P\left(Z > \frac{208 - 190}{5/\sqrt{25}}\right) = P(Z > 2.57) = 0.0051$$

5.27. (a) The mean of six untreated specimens has a standard deviation of $2.2/\sqrt{6} \doteq 0.8981$ lbs, so $P(\bar{x}_u > 50) = P\left(Z > \frac{50 - 57}{0.8981}\right) = P(Z > -7.79)$, which is basically 1. **(b)** $\bar{x}_u - \bar{x}_t$ has mean $57 - 30 = 27$ lbs and standard deviation $\sqrt{2.2^2/6 + 1.6^2/6} \doteq 1.1106$ lbs, so $P(\bar{x}_u - \bar{x}_t > 25) = P\left(Z > \frac{25 - 27}{1.1106}\right) = P(Z > -1.80) \doteq 0.9641$.

5.29. (a) $\bar{y}$ has a $N(\mu_Y, \sigma_Y/\sqrt{m})$ distribution and $\bar{x}$ has a $N(\mu_X, \sigma_X/\sqrt{n})$ distribution. **(b)** $\bar{y} - \bar{x}$ has a Normal distribution with mean $\mu_Y - \mu_X$ and standard deviation $\sqrt{\sigma_Y^2/m + \sigma_X^2/n}$.

5.31. The total height H of the four rows has a Normal distribution with mean $4 \times 8 = 32$ inches and standard deviation $0.1\sqrt{4} = 0.2$ inch. $P(H < 31.5 \text{ or } H > 32.5) = 1 - P(31.5 < H < 32.5) = 1 - P(-2.50 < Z < 2.50) = 1 - 0.9876 = 0.0124$.

5.33. (a) $n = 1500$ (the sample size). **(b)** The "Yes" count seems like the most reasonable choice, but either count is defensible. **(c)** $X = 825$ (or $X = 675$). **(d)** $\hat{p} = \frac{825}{1500} = 0.55$ (or $\hat{p} = \frac{675}{1500} = 0.45$).

5.35. We have 15 independent trials, each with probability of success (heads) equal to 0.5, so X has the $B(15, 0.5)$ distribution.

5.37. (a) For the $B(5, 0.4)$ distribution, $P(X = 0) = 0.0778$ and $P(X \geq 3) = 0.3174$. **(b)** For the $B(5, 0.6)$ distribution, $P(X = 5) = 0.0778$ and $P(X \leq 2) = 0.3174$. **(c)** The number of "failures" in the $B(5, 0.4)$ distribution has the $B(5, 0.6)$ distribution. With 5 trials, 0 successes is equivalent to 5 failures, and 3 or more successes is equivalent to 2 or fewer failures.

5.39. (a) $\hat{p}$ has approximately a Normal distribution with mean 0.5 and standard deviation 0.05, so $P(0.3 < \hat{p} < 0.7) = P(-4 < Z < 4) \doteq 1$. **(b)** $P(0.35 < \hat{p} < 0.65) = P(-3 < Z < 3) \doteq 0.9974$.

　　Note: *For the second, the 68–95–99.7 rule would give 0.997 — an acceptable answer, especially since this is an approximation anyway. For comparison, the exact answers (to four decimal places) are $P(0.3 < \hat{p} < 0.7) \doteq 0.9999$ or $P(0.3 \leq \hat{p} \leq 0.7) \doteq 1.0000$, and $P(0.35 < \hat{p} < 0.65) \doteq 0.9965$ or $P(0.35 \leq \hat{p} \leq 0.65) \doteq 0.9982$. (Notice that the "correct" answer depends on our understanding of "between.")*

5.41. (a) Separate flips are independent (coins have no "memory," so they do not try to compensate for a lack of tails). **(b)** Separate flips are independent (coins have no "memory," so they do not get on a "streak" of heads). **(c)** $\hat{p}$ can vary from one set of observed data to another; it is not a parameter.

5.43. (a) A $B(200, p)$ distribution seems reasonable for this setting (even though we do not know what p is). **(b)** This setting is not binomial; there is no fixed value of n. **(c)** A $B(500, 1/12)$ distribution seems appropriate for this setting. **(d)** This is not binomial, because separate cards are not independent.

5.45. (a) C, the number caught, is $B(10, 0.7)$. M, the number missed, is $B(10, 0.3)$. **(b)** Referring to Table C, we find $P(M \geq 4) = 0.2001 + 0.1029 + 0.0368 + 0.0090 + 0.0014 + 0.0001 = 0.3503$ (software: 0.3504).

5.47. (a) The mean of C is $(10)(0.7) = 7$ errors caught; for M the mean is $(10)(0.3) = 3$ errors missed. **(b)** The standard deviation of C (or M) is $\sigma = \sqrt{(10)(0.7)(0.3)} \doteq 1.4491$ errors. **(c)** With $p = 0.9$, $\sigma = \sqrt{(10)(0.9)(0.1)} \doteq 0.9487$ errors; with $p = 0.99$, $\sigma \doteq 0.3146$ errors. σ decreases toward 0 as p approaches 1.

5.49. $m = 6$: $P(X \geq 6) = 0.0473$ and $P(X \geq 5) = 0.1503$.

5.51. The count of 5s among n random digits has a binomial distribution with $p = 0.1$.
(a) $P(\text{at least one } 5) = 1 - P(\text{no } 5) = 1 - (0.9)^5 \doteq 0.4095$. (Or take 0.5905 from Table C and subtract from 1.) **(b)** $\mu = (40)(0.1) = 4$.

5.53. (a) $n = 4$ and $p = 1/4 = 0.25$. **(b)** The distribution is below; the histogram is on the right. **(c)** $\mu = np = 1$.

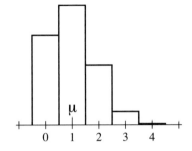

x	0	1	2	3	4
$P(X = x)$	.3164	.4219	.2109	.0469	.0039

5.55. Recall that $\hat{p}$ is approximately Normally distributed with mean $\mu = p$ and standard deviation $\sqrt{p(1-p)/n}$. **(a)** With $p = 0.24$, $\sigma \doteq 0.01333$, so $P(0.22 < \hat{p} < 0.26) = P(-1.50 < Z < 1.50) \doteq 0.8664$. (Software computation of the Normal probability gives 0.8666. Using a binomial distribution, we can also find $P(226 \le X \le 267) \doteq 0.8752$.) **(b)** With $p = 0.04$, $\sigma \doteq 0.00611$, so $P(0.02 < \hat{p} < 0.06) = P(-3.27 < Z < 3.27) = 0.9990$. (Using a binomial distribution, we can also find $P(21 \le X \le 62) \doteq 0.9992$.) **(c)** $P(-0.02 < \hat{p} - p < 0.02)$ increases to 1 as p gets closer to 0. (This is because σ also gets close to 0, so that $0.02/\sigma$ grows.)

5.57. (a) The mean is $\mu = p = 0.69$, and the standard deviation is $\sigma = \sqrt{p(1-p)/n} \doteq 0.0008444$. **(b)** $\mu \pm 2\sigma$ gives the range 68.83% to 69.17%. **(c)** This range is considerably narrower than the historical range. In fact, 67% and 70% correspond to $z = -23.7$ and $z = 11.8$—suggesting that the observed percents do not come from a $N(0.69, 0.0008444)$ distribution; that is, the population proportion has changed over time.

5.59. (a) $\hat{p} = \frac{140}{200} = 0.7$. **(b)** We want $P(X \ge 140)$ or $P(\hat{p} \ge 0.7)$. The first can be found exactly (using a binomial distribution), or we can compute either

			Continuity correction	
Exact prob.	Table Normal	Software Normal	Table Normal	Software Normal
0.2049	0.1841	0.1835	0.2033	0.2041

using a Normal approximation (with or without the continuity correction). All possible answers are shown on the right. **(c)** The sample results are higher than the national percentage, but the sample was so small that such a difference could arise by chance even if the true campus proportion is the same.

5.61. (a) $p = 1/4 = 0.25$. **(b)** $P(X \ge 10) = 0.0139$. **(c)** $\mu = np = 5$ and $\sigma = \sqrt{np(1-p)} = \sqrt{3.75} \doteq 1.9365$ successes. **(d)** No: The trials would not be independent because the subject may alter his/her guessing strategy based on this information.

5.63. (a) X, the count of successes, has the $B(900, 1/5)$ distribution, with mean $\mu_X = np = (900)(1/5) = 180$ and $\sigma_X = \sqrt{(900)(0.2)(0.8)} = 12$ successes. **(b)** For $\hat{p}$, the mean is $\mu_{\hat{p}} = p = 0.2$ and $\sigma_{\hat{p}} = \sqrt{(0.2)(0.8)/900} \doteq 0.01333$. **(c)** $P(\hat{p} > 0.24) \doteq P(Z > 3) = 0.0013$. **(d)** From a standard Normal distribution,

$P(Z > 2.326) = 0.01$, so the subject must score 2.326 standard deviations above the mean: $\mu_{\hat{p}} + 2.326\sigma_{\hat{p}} = 0.2310$. This corresponds to 208 or more successes.

5.65. (a) $p = \frac{23,772,494}{209,128,094} \doteq 0.1137$. **(b)** If B is the number of blacks, then B has (approximately) the $B(1200, 0.1137)$ distribution, so the mean is $np \doteq 136.4$ blacks. **(c)** $P(B \le 100) \doteq P(Z < -3.31) = 0.0005$.

    Note: *In (b), the population is at least 20 times as large as the sample, so our "rule of thumb" for using a binomial distribution is satisfied. In fact, the mean would be the same even if we could not use a binomial distribution, but we need to have a binomial distribution for part (c), so that we can approximate it with a Normal distribution—which we can safely do, because both np and n(1 − p) are much greater than 10.*

5.67. Jodi's number of correct answers will have the $B(n, 0.88)$ distribution. **(a)** $P(\hat{p} \le 0.85) = P(X \le 85)$ is on line 1. **(b)** $P(\hat{p} \le 0.85) = P(X \le 212)$ is on line 2. **(c)** For a test with 400

Exact prob.	Table Normal	Software Normal	Continuity correction	
			Table Normal	Software Normal
0.2160	0.1788	0.1780	0.2206	0.2209
0.0755	0.0594	0.0597	0.0721	0.0722

questions, the standard deviation of $\hat{p}$ would be half as big as the standard deviation of $\hat{p}$ for a test with 100 questions: With $n = 100$, $\sigma = \sqrt{(0.88)(0.12)/100} \doteq 0.03250$; and with $n = 400$, $\sigma = \sqrt{(0.88)(0.12)/400} \doteq 0.01625$. **(d)** Yes: Regardless of p, n must be quadrupled to cut the standard deviation in half.

5.69. Y has possible values $1, 2, 3, \ldots$. $P(\text{first } \boxdot \text{ appears on toss } k) = \left(\frac{5}{6}\right)^{k-1}\left(\frac{1}{6}\right)$.

5.71. (a) With $\sigma_{\bar{x}} = 0.08/\sqrt{3} \doteq 0.04619$, $\bar{x}$ has (approximately) a $N(123 \text{ mg}, 0.04619 \text{ mg})$ distribution. **(b)** $P(\bar{x} \ge 124) = P(Z \ge 21.65)$, which is essentially 0.

5.73. (a) Out of 12 independent vehicles, the number X with one person has the $B(12, 0.755)$ distribution, so $P(X \ge 7) = 0.9503$ (using software or a calculator). **(b)** Y (the number of one-person cars in a sample of 80) has the $B(80, 0.755)$ distribution. Regardless of the approach used—Normal approximation, or exact computation using software or a calculator—$P(Y \ge 41) \doteq 1$.

5.75. The probability that the first digit is 1, 2, or 3 is $0.301 + 0.176 + 0.125 = 0.602$, so the number of invoices for amounts beginning with these digits should have a binomial distribution

Table Normal	Software Normal	Continuity correction	
		Table Normal	Software Normal
0.0034	0.0033	0.0037	0.0037

with $n = 1000$ and $p = 0.602$. More usefully, the proportion $\hat{p}$ of such invoices should have approximately a Normal distribution with mean $p = 0.602$ and standard deviation $\sqrt{p(1-p)/1000} \doteq 0.01548$, so $P(\hat{p} \le \frac{560}{1000}) \doteq P(Z \le -2.71) = 0.0034$. Alternate answers shown on the right.

5.77. If $\bar{x}$ is the average weight of 12 eggs, then $\bar{x}$ has a $N(65 \text{ g}, 5/\sqrt{12} \text{ g}) = N(65 \text{ g}, 1.4434 \text{ g})$ distribution, and $P(\frac{755}{12} < \bar{x} < \frac{830}{12}) \doteq P(-1.44 < Z < 2.89) = 0.9231$ (software: 0.9236).

5.79. The center line is $\mu_{\bar{x}} = \mu = 4.25$ and the control limits are $\mu \pm 3\sigma/\sqrt{5} = 4.0689$ to 4.4311.

5.81. (a) $\hat{p}_F$ is approximately $N(0.82, 0.01921)$ and $\hat{p}_M$ is approximately $N(0.88, 0.01625)$. **(b)** When we subtract two independent Normal random variables, the difference is Normal. The new mean is the difference of the two means ($0.88 - 0.82 = 0.06$), and the new variance is the sum of the variances ($0.01921^2 + 0.01625^2 = 0.000633$), so $\hat{p}_M - \hat{p}_F$ is approximately $N(0.06, 0.02516)$. **(c)** $P(\hat{p}_F > \hat{p}_M) = P(\hat{p}_M - \hat{p}_F < 0) \doteq P(Z < -2.38) = 0.0087$ (software: 0.0085).

5.83. For each step of the random walk, the mean is $\mu = (1)(0.6) + (-1)(0.4) = 0.2$, the variance is $\sigma^2 = (1 - 0.2)^2(0.6) + (-1 - 0.2)^2(0.4) = 0.96$, and the standard deviation is $\sigma = \sqrt{0.96} \doteq 0.9798$. Therefore, $Y/500$ has approximately a $N(0.2, 0.04382)$ distribution, and $P(Y \geq 200) = P(\frac{Y}{500} \geq 0.4) \doteq P(Z \geq 4.56) \doteq 0$.

   Note: *The number R of right-steps has a binomial distribution with n = 500 and p = 0.6. Y $\geq$ 200 is equivalent to taking at least 350 right-steps, so we can also compute this probability as P(R $\geq$ 350), for which software gives the exact value 0.00000215....*

Chapter 6 Solutions

6.1. $\sigma_{\bar{x}} = \sigma/\sqrt{25} = \frac{\$2.50}{5} = \$0.50.$

6.3. As in the previous solution, take two standard deviations: $1.00.
   Note: *This is the whole idea behind a confidence interval: Probability tells us that $\bar{x}$ is usually close to μ. That is equivalent to saying that μ is usually close to $\bar{x}$.*

6.5. The standard error is $s_{\bar{x}} = \frac{\sigma}{\sqrt{100}} = 0.3$, and the 95% confidence interval for μ is

$$87.3 \pm 1.96 \left(\frac{3}{\sqrt{100}} \right) = 87.3 \pm 0.588 = 86.712 \text{ to } 87.888$$

6.7. $n = \left(\frac{(1.96)(12,000)}{1000} \right)^2 \doteq 553.19$—take $n = 554$.

6.9. The (useful) response rate is $\frac{249}{5800} \doteq 0.0429$, or about 4.3%. The reported margin of error is probably unreliable because we know nothing about the 95.7% of students that did *not* provide (useful) responses; they may be more (or less) likely to charge education-related expenses.

6.11. The margins of error are $1.96 \times 7/\sqrt{n}$, which yields 4.3386, 3.0679, 2.1693, and 1.5339. (And, of course, all intervals are centered at 50.) Interval width decreases with increasing sample size.

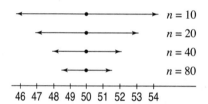

6.13. **(a)** She did not divide the standard deviation by $\sqrt{n} = 20$. **(b)** Confidence intervals concern the population mean, not the sample mean. (The value of the sample mean is known to be 8.6; it is the population mean that we do not know.) **(c)** 95% is a confidence level, not a probability. Furthermore, it does not make sense to make probability statements about the population mean μ, which is an unknown constant (rather than a random quantity). **(d)** The large sample size does not affect the distribution of individual alumni ratings (the population distribution). The use of a Normal distribution is justified because the distribution *of the sample mean* is approximately Normal when the sample is large.
   Note: *For part (c), a Bayesian statistician might view the population mean μ as a random quantity, but the viewpoint taken in the text is non-Bayesian.*

6.15. **(a)** To estimate the mean importance of recreation to college satisfaction, the 95% confidence interval for μ is

$$7.5 \pm 1.96 \left(\frac{3.9}{\sqrt{2673}} \right) = 7.5 \pm 0.1478 = 7.3522 \text{ to } 7.6478$$

(b) The 99% confidence interval for μ is

$$7.5 \pm 2.576 \left(\frac{3.9}{\sqrt{2673}} \right) = 7.5 \pm 0.1943 = 7.3057 \text{ to } 7.6943$$

6.17. For mean TRAP level, the margin of error is 2.29 U/l and the 95% confidence interval for μ is

$$13.2 \pm 1.96 \left(\frac{6.5}{\sqrt{31}} \right) = 13.2 \pm 2.29 = 10.91 \text{ to } 15.49 \text{ U/l}$$

6.19. Scenario B has a smaller margin of error. Both samples would have the same value of z^* (1.96), but the value of σ would be smaller for (B) because we would have less variability in textbook cost for students in a single major.

   Note: *Of course, at some schools, taking a sample of 100 sophomores in a given major is not possible. However, even if we sampled students from a number of institutions, we still might expect less variability within a given major than from a broader cross-section.*

6.21. (a) "The probability is about 0.95 that $\bar{x}$ is within 14 kcal/day of ... μ" (because 14 is two standard deviations). **(b)** This is simply another way of understanding the statement from part (a): If $|\bar{x} - \mu|$ is less than 14 kcal/day 95% of the time, then "about 95% of all samples will capture the true mean ... in the interval $\bar{x}$ plus or minus 14 kcal/day."

6.23. No; This is a range of values for the mean rent, not for individual rents.

   Note: *To find a range to include 95% of all rents, we should take $\mu \pm 2\sigma$ (or more precisely, $\mu \pm 1.96\sigma$), where μ is the (unknown) mean rent for all apartments, and σ is the standard deviation for all apartments (assumed to be \$290 in Exercise 6.22). If μ were equal to \$1050, for example, this range would be about \$470 to \$1630. However, because we do not actually know μ, we estimate it using $\bar{x}$, and to account for the variability in $\bar{x}$, we must widen the margin of error by a factor of $\sqrt{1 + \frac{1}{n}}$. The formula $\bar{x} \pm 2\sigma\sqrt{1 + \frac{1}{10}}$ is called a* prediction interval *for future observations. (Usually, such intervals are constructed with the t distribution, discussed in the Chapter 7, but the idea is the same.)*

6.25. (a) For the mean number of hours spent on the Internet, the 95% confidence interval for μ is

$$19 \pm 1.96 \left(\frac{5.5}{\sqrt{1200}} \right) = 19 \pm 0.3112 = 18.6888 \text{ to } 19.3112 \text{ hours}$$

(b) No; this is a range of values for the mean time spent, not for individual times. (See also the comment in the solution to Exercise 6.23.)

6.27. (a) We can be 95% confident, but not *certain*. **(b)** We obtained the interval 85% to 95% by a method that gives a correct result (that is, includes the true mean) 95% of the time. **(c)** For 95% confidence, the margin of error is about two standard deviations (that is, $z^* = 1.96$), so $\sigma_{\text{estimate}} \doteq 2.5\%$. **(d)** No; confidence intervals only account for random sampling error.

6.29. Multiply by $\dfrac{1.609 \text{ km}}{1 \text{ mile}} \cdot \dfrac{1 \text{ gallon}}{3.785 \text{ liters}} \doteq 0.4251 \dfrac{\text{kpl}}{\text{mpg}}$. This gives $\bar{x}_{\text{kpl}} = 0.4251\bar{x}_{\text{mpg}} \doteq 18.3515$ and margin of error $1.96(0.4251\sigma_{\text{mpg}})/\sqrt{20} \doteq 0.6521$ kpl, so the 95% confidence interval is 17.6994 to 19.0036 kpl.

6.31. $n = \left(\dfrac{(1.96)(6.5)}{1.5} \right)^2 \doteq 72.14$—take $n = 73$.

6.33. No: Because the numbers are based on voluntary response rather than an SRS, the confidence interval methods of this chapter cannot be used; the interval does not apply to the whole population.

6.35. The number of hits has a binomial distribution with parameters $n = 5$ and $p = 0.95$, so the number of misses is binomial with $n = 5$ and $p = 0.05$. We can therefore use Table C to answer these questions. **(a)** The probability that all cover their means is $0.95^5 \doteq 0.7738$. (Or use Table C to find the probability of 0 misses.) **(b)** The probability that at least four intervals cover their means is $0.95^5 + 5(0.05)(0.95^4) \doteq 0.9774$. (Or use Table C to find the probability of 0 or 1 misses.)

6.37. If μ is the mean DXA reading for the phantom, we test H_0: $\mu = 1.4$ g/cm^2 versus H_a: $\mu \neq 1.4$ g/cm^2.

6.39. $P(Z < -1.63) = 0.0516$, so the two-sided P-value is $2(0.0516) = 0.1032$.

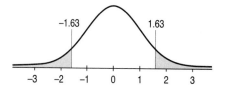

6.41. **(a)** For $P = 0.05$, the value of z is 1.645. **(b)** For a one-sided alternative (on the positive side), z is statistically significant at $\alpha = 0.05$ if $z > 1.645$.

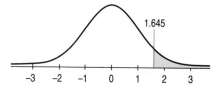

6.43. **(a)** $z = \frac{27 - 25}{5/\sqrt{36}} = 2.4$. **(b)** For a one-sided alternative, $P = P(Z > 2.4) = 0.0082$. **(c)** For a two-sided alternative, double the one-sided P-value: $P = 0.0164$.

6.45. Recall the statement from the text: "A level α two-sided significance test rejects ... H_0: $\mu = \mu_0$ exactly when the value μ_0 falls outside a level $1 - \alpha$ confidence interval for μ." **(a)** No; 30 is not in the 95% confidence interval because $P = 0.033$ means that we would reject H_0 at $\alpha = 0.05$. **(b)** Yes; 30 is in the 99% confidence interval because we would *not* reject H_0 at $\alpha = 0.01$.

6.47. **(a)** Yes, we reject H_0 at $\alpha = 0.05$. **(b)** No, we do not reject H_0 at $\alpha = 0.01$. **(c)** We have $P = 0.039$; we reject H_0 at significance level α if $P < \alpha$.

6.49. **(a)** One of the one-sided P-values is half as big as the two-sided P-value (0.022); the other is $1 - 0.022 = 0.978$. **(b)** Suppose the null hypothesis is H_0: $\mu = \mu_0$. The smaller P-value (0.022) goes with the one-sided alternative that is consistent with the observed data; for example, if $\bar{x} > \mu_0$, then $P = 0.022$ for the alternative $\mu > \mu_0$.

6.51. **(a)** Hypotheses should be stated in terms of the population mean, not the sample mean. **(b)** The null hypothesis H_0 should be that there is no change ($\mu = 21.2$). **(c)** A small P-value is needed for significance; $P = 0.98$ gives no reason to reject H_0. **(d)** We compare the P-value, not the z-statistic, to α. (In this case, such a small value of z would have a very large *P*-value—close to 0.5 for a one-sided alternative, or close to 1 for a two-sided alternative.)

6.53. (a) If μ is the mean score for the population of placement-test students, then we test H_0: $\mu = 77$ versus H_a: $\mu \neq 77$ because we have no prior belief about whether placement-test students will do better or worse. **(b)** If μ is the mean time to complete the maze with rap music playing, then we test H_0: $\mu = 20$ seconds versus H_a: $\mu > 20$ seconds because we believe rap music will make the mice finish more slowly. **(c)** If μ is the mean area of the apartments, we test H_0: $\mu = 880$ ft^2 versus H_a: $\mu < 880$ ft^2, because we suspect the apartments are smaller than advertised.

6.55. (a) H_0: $\mu = \$42,800$ versus H_a: $\mu > \$42,800$, where μ is the mean household income of mall shoppers. **(b)** H_0: $\mu = 0.4$ hr versus H_a: $\mu \neq 0.4$ hr, where μ is this year's mean response time.

6.57. (a) For H_a: $\mu > \mu_0$, the *P*-value is $P(Z > -1.82) = 0.9656$.
(b) For H_a: $\mu < \mu_0$, the *P*-value is $P(Z < -1.82) = 0.0344$.
(c) For H_a: $\mu \neq \mu_0$, the *P*-value is $2P(Z < -1.82) = 2(0.0344) = 0.0688$.

6.59. Recall the statement from the text: "A level α two-sided significance test rejects ... H_0: $\mu = \mu_0$ exactly when the value μ_0 falls outside a level $1 - \alpha$ confidence interval for μ." **(b)** Yes, we would reject H_0: $\mu = 24$; the fact that 24 falls outside the 90% confidence interval means that $P < 0.10$. **(a)** No, we would not reject H_0: $\mu = 30$ because 30 falls inside the confidence interval, so $P > 0.10$.

   Note: *The given confidence interval suggests that $\bar{x} = 28.5$, and if the interval was constructed using the Normal distribution, the standard error of the mean is about 1.75—half the margin of error. (The standard error might be less if it was constructed with a t distribution rather than the Normal distribution.) Then $\bar{x}$ is about 2.57 standard errors above $\mu = 24$—yielding $P \doteq 0.01$—and $\bar{x}$ is about 0.86 standard errors below $\mu = 30$, so that $P \doteq 0.39$.*

6.61. $P = 0.09$ means there is some evidence for the wage decrease, but it is not significant at the $\alpha = 0.05$ level. Specifically, the researchers observed that average wages for peer-driven students were 13% lower than average wages for ability-driven students, but (when considering overall variation in wages) such a difference might arise by chance 9% of the time, even if student motivation had no effect on wages.

6.63. Even if the two groups (the health and safety class, and the statistics class) had the same level of alcohol awareness, there might be some difference in our sample due to chance. The difference observed was large enough that it would rarely arise by chance. The reason for this difference might be that health issues related to alcohol use are probably discussed in the health and safety class.

6.65. If μ is the mean difference between the two groups of children, we test H_0: $\mu = 0$ versus H_a: $\mu \neq 0$. The test statistic is $z = \frac{4-0}{1.2} \doteq 3.33$, for which software reports $P \doteq 0.0009$—very strong evidence against the null hypothesis.

Note: *The exercise reports the standard deviation of the mean, rather than the sample standard deviation; that is, the reported value has already been divided by $\sqrt{238}$.*

6.67. If μ is the mean east-west location, the hypotheses are H_0: $\mu = 100$ versus H_a: $\mu \neq 100$ (as in the previous exercise). For testing these hypotheses, we find $z = \frac{113.8 - 100}{58/\sqrt{584}} \doteq 5.75$. This is highly significant ($P < 0.0001$), so we conclude that the trees are not uniformly spread from east to west.

6.69. (a) $z = \frac{127.8 - 115}{30/\sqrt{25}} \doteq 2.13$, so the P-value is $P = P(Z > 2.13) = 0.0166$. This is strong evidence that the older students have a higher SSHA mean. **(b)** The important assumption is that this is an SRS from the population of older students. We also assume a Normal distribution, but this is not crucial provided there are no outliers and little skewness.

6.71. (a) H_0: $\mu = 0$ mpg versus H_a: $\mu \neq 0$ mpg, where μ is the mean difference. **(b)** The mean of the 20 differences is $\bar{x} = 2.73$, so $z = \frac{2.73 - 0}{3/\sqrt{20}} \doteq 4.07$, for which $P < 0.0001$. We conclude that $\mu \neq 0$ mpg; that is, we have strong evidence that the computer's reported fuel efficiency differs from the driver's computed values.

6.73. For (b) and (c), either compare with the critical values in Table D or determine the P-value (0.0336). **(a)** H_0: $\mu = 0.9$ mg versus H_a: $\mu > 0.9$ mg. **(b)** Yes, because $z > 1.645$ (or because $P < 0.05$). **(c)** No, because $z < 2.326$ (or because $P > 0.01$).

6.75. See the sample screen (for $\bar{x} = 1$) on the right. As one can judge from the shading under the Normal curve, $\bar{x} = 0.7$ is not significant, but 0.8 is. (In fact, the cutoff is about 0.7354, which is approximately $2.326/\sqrt{10}$.) Smaller α means that $\bar{x}$ must be farther away from μ_0 in order to reject H_0.

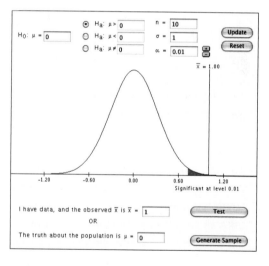

6.77. When a test is significant at the 5% level, it means that if the null hypothesis were true, outcomes similar to those seen are expected to occur fewer than 5 times in 100 repetitions of the experiment or sampling. "Significant at the 10% level" means we have observed something that occurs in fewer than 10 out of 100 repetitions (when H_0 is true). Something that occurs "fewer than 5 times in 100 repetitions" also occurs "fewer than 10 times in 100 repetitions," so significance at the 5% level implies significance at the 10% level (or any higher level).

6.79. Using Table D or software, we find that the 0.005 critical value is 2.576, and the 0.0025 critical value is 2.807. Therefore, if $2.576 < |z| < 2.807$—that is, either $2.576 < z < 2.807$ or $-2.807 < z < -2.576$—then z would be significant at the 1% level, but not at the 0.5% level.

6.81. As $0.63 < 0.674$, the one-sided P-value is $P > 0.25$. (Software gives $P = 0.2643$.)

6.83. Because the alternative is two-sided, the answer for $z = -1.92$ is the same as for $z = 1.92$: $-1.645 > -1.92 > -1.960$, so Table D says that $0.05 < P < 0.10$, and Table A gives $P = 2(0.0274) = 0.0548$.

6.85. In order to determine the effectiveness of alarm systems, we need to know the percent of all homes with alarm systems, and the percent of burglarized homes with alarm systems. For example, if only 10% of all home have alarm systems, then we should compare the proportion of burglarized homes with alarm systems to 10%, not 50%.

   An alternate (but rather impractical) method would be to sample homes and classify them according to whether or not they had an alarm system, and also by whether or not they had experienced a break-in at some point in the recent past. This would likely require a very large sample in order to get a sufficiently large count of homes that had experienced break-ins.

6.87. The first test was barely significant at $\alpha = 0.05$, while the second was significant at any reasonable α.

6.89. A significance test answers only Question b. The P-value states how likely the observed effect (or a stronger one) is if H_0 is true, and chance alone accounts for deviations from what we expect. The observed effect may be significant (very unlikely to be due to chance) and yet not be of practical importance. And the calculation leading to significance *assumes* a properly designed study.

6.91. (a) If SES had no effect on LSAT results, there would still be some difference in scores due to chance variation. "Statistically insignificant" means that the observed difference was no more than we might expect from that chance variation. (b) If the results are based on a small sample, then even if the null hypothesis were not true, the test might not be sensitive enough to detect the effect. Knowing the effects were small tells us that the statistically insignificant test result did not occur merely because of a small sample size.

6.93. In each case, we find the test statistic z by dividing the observed difference ($2453.7 - 2403.7 = 50$ kcal/day) by $880/\sqrt{n}$. (a) For $n = 100$, $z \doteq 0.57$, so $P = P(Z > 0.57) = 0.2843$. (b) For $n = 500$, $z \doteq 1.27$, so $P = P(Z > 1.27) = 0.1020$. (c) For $n = 2500$, $z \doteq 2.84$, so $P = P(Z > 2.84) = 0.0023$.

6.95. We expect more variation with small sample sizes, so even a large difference between $\bar{x}$ and μ_0 (or whatever measures are appropriate in our hypothesis test) might not turn out to be significant. If we were to repeat the test with a larger sample, the decrease in the standard error might give us a small enough P-value to reject H_0.

6.101. $P = 0.00001 = \frac{1}{100,000}$, so we would need $n = 100{,}000$ tests in order to expect one P-value of this size (assuming that all null hypotheses are true). That is why we reject H_0 when we see P-values such as this: It indicates that our results would rarely happen if H_0 were true.

6.103. Using $\alpha/12 \doteq 0.004167$ as the cutoff, we reject the fifth ($P = 0.002$) and eleventh ($P < 0.002$) tests.

6.105. A larger sample gives more information and therefore gives a better chance of detecting a given alternative; that is, larger samples give more power.

6.107. The power for $\mu = 40$ will be higher than 0.6, because larger differences are easier to detect. The picture on the right shows one way to illustrate this (assuming Normal distributions): The solid curve (centered at 20) is the distribution under

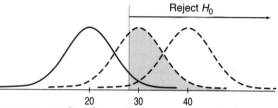

the null hypothesis, and the two dashed curves represent the alternatives $\mu = 30$ and $\mu = 40$. The shaded region under the middle curve is the power against $\mu = 30$; that is, that shaded region is 60% of the area under that curve. The power against $\mu = 40$ would be the corresponding area under the rightmost curve, which would clearly be greater than 0.6.

6.109. The applet reports the power as 0.986.

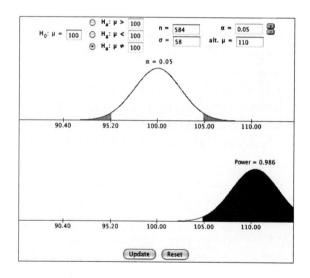

6.111. We reject H_0 when $z > 2.326$, which is equivalent to $\bar{x} > 450 + 2.326 \cdot \frac{100}{\sqrt{500}} \doteq 460.4$, so the power against $\mu = 460$ is

$$P(\text{reject } H_0 \text{ when } \mu = 460) = P(\bar{x} > 460.4 \text{ when } \mu = 460)$$

$$= P\left(Z > \frac{460.4 - 460}{100/\sqrt{500}}\right) \doteq P(Z > 0.09) = 0.4641.$$

This is quite a bit less than the "80% power" standard.

6.113. **(a)** The hypotheses are "subject should go to college" and "subject should join work force." The two types of errors are recommending someone go to college when (s)he is better suited for the work force, and recommending the work force for someone who should go to college. **(b)** In significance testing, we typically wish to decrease the probability of wrongly rejecting H_0 (that is, we want α to be small); the answer to this question depends on which hypothesis is viewed as H_0.

   Note: *For part (a), there is no clear choice for which should be the null hypothesis. In the past, when fewer people went to college, one might have chosen "work force" as H_0—that is, one might have said, "we'll assume this student will join the work force unless we are convinced otherwise." Presently, roughly two-thirds of graduates attend college, which might suggest H_0 should be "college."*

6.115. From the description, we might surmise that we had two (or more) groups of students—say, an exercise group and a control (or no-exercise) group. **(a)** For example, if μ is the mean difference in scores between the two groups, we might test H_0: $\mu = 0$ versus H_a: $\mu \neq 0$. (Assuming we had no prior suspicion about the effect of exercise, the alternative should be two-sided.) **(b)** With $P = 0.38$, we would not reject H_0. In plain language: The results observed do not differ greatly from what we would expect if exercise had no effect on exam scores. **(c)** For example: Was this an experiment? What was the design? How big were the samples?

6.117. **(a)** Because all standard deviations and sample sizes are the same, the margin of error for all intervals is $1.96 \times 19/\sqrt{180} \doteq 2.7757$. The confidence intervals are listed in the table below. **(b)** The plot below shows the error bars for the confidence intervals of (a), and also for part (c). The limits for (a) are the thicker lines which do not extend as far above and below the mean. **(c)** With $z^* = 2.40$, the margin of error for all intervals is $2.40 \times 19/\sqrt{180} \doteq 3.3988$. The confidence intervals are listed in the table below and are shown in the plot (the thinner lines with the wider dashes). **(d)** When we use $z^* = 2.40$ to adjust for the fact that we are making three "simultaneous" confidence intervals, the margin of error is larger, so the intervals overlap more.

Workplace size	Mean SCI
< 50	64.45 to 70.01
50–200	67.59 to 73.15
> 200	72.05 to 77.61
< 50	63.83 to 70.63
50–200	66.97 to 73.77
> 200	71.43 to 78.23

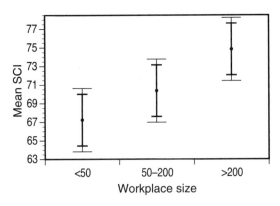

6.119. A sample screenshot and example plot are not shown but would be similar to those shown on the following page (from the solution to Exercise 6.118). Most students (99.4% of them) should find that their final proportion is between 0.84 and 0.96; 85% will have a proportion between 0.87 and 0.93.

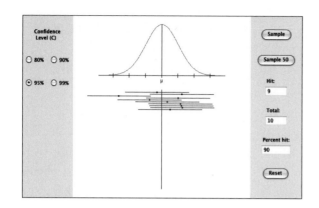 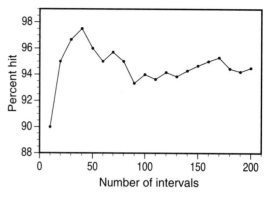

6.121. (a) $\bar{x} = 5.\overline{3}$ mg/dl, so $\bar{x} \pm 1.960\sigma/\sqrt{6}$ is 4.6132 to 6.0534 mg/dl. **(b)** To test H_0: $\mu = 4.8$ mg/dl versus H_a: $\mu > 4.8$ mg/dl, we compute $z = \frac{\bar{x}-4.8}{0.9/\sqrt{6}} \doteq 1.45$ and $P \doteq 0.0735$. This is not strong enough to reject H_0.

   Note: *The confidence interval in (a) would allow us to say without further computation that, against a two-sided alternative, we would have P > 0.05. Because we have a one-sided alternative, we could conclude from the confidence interval that P > 0.025, but that is not enough information to draw a conclusion.*

6.123. (a) The stemplot is reasonably symmetric for such a small sample.
   (b) $\bar{x} = 30.4$ µg/l; $30.4 \pm (1.96)(7/\sqrt{10})$ gives 26.0614 to 34.7386 µg/l.
   (c) We test H_0: $\mu = 25$ µg/l versus H_a: $\mu > 25$ µg/l. $z = \frac{30.4-25}{7/\sqrt{10}} \doteq 2.44$, so $P = 0.0073$. (We knew from (b) that it had to be smaller than 0.025). This is fairly strong evidence against H_0; the beginners' mean threshold is higher than 25 µg/l.

```
2 | 034
2 |
3 | 01124
3 | 6
4 | 3
```

6.125. (a) Under H_0, $\bar{x}$ has a $N(0\%, 55\%/\sqrt{104}) \doteq N(0\%, 5.3932\%)$ distribution. **(b)** $z = \frac{6.9-0}{55/\sqrt{104}} \doteq 1.28$, so $P = P(Z > 1.28) = 0.1003$. **(c)** This is not significant at $\alpha = 0.05$. The study gives *some* evidence of increased compensation, but it is not very strong; similar results would happen about 10% of the time just by chance.

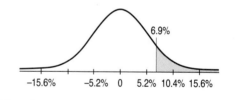

6.127. Yes. That's the heart of why we care about statistical significance. Significance tests allow us to discriminate between random differences ("chance variation") that might occur when the null hypothesis is true, and differences that are unlikely to occur when H_0 is true.

6.129. For each sample, find $\bar{x}$, then take $\bar{x} \pm 1.96(4/\sqrt{12}) = \bar{x} \pm 2.2632$.
   We "expect" to see that 95 of the 100 intervals will include 25 (the true value of μ); binomial computations show that (about 99% of the time) 90 or more of the 100 intervals will include 20.

6.131. For each sample, find $\bar{x}$, then compute $z = \frac{\bar{x}-23}{4/\sqrt{12}}$. Choose a significance level α and the appropriate cutoff point (z^*)—for example, with $\alpha = 0.10$, reject H_0 if $|z| > 1.645$; with $\alpha = 0.05$, reject H_0 if $|z| > 1.96$.

Because the true mean is 25, $Z = \frac{\bar{x}-25}{4/\sqrt{12}}$ has a $N(0, 1)$ distribution, so the probability that we will accept H_0 is $P\left(-z^* < \frac{\bar{x}-23}{4/\sqrt{12}} < z^*\right) = P(-z^* < Z + 1.7321 < z^*) = P(-1.7321 - z^* < Z < -1.7321 + z^*)$. If $\alpha = 0.10$ ($z^* = 1.645$), this probability is $P(-3.38 < Z < -0.09) = 0.4637$; if $\alpha = 0.05$ ($z^* = 1.96$), this probability is $P(-3.69 < Z < 0.23) = 0.5909$. For smaller α, the probability will be larger. Thus we "expect" to (wrongly) accept H_0 about half the time (or more), and correctly reject H_0 about half the time or less. (The probability of rejecting H_0 is essentially the power of the test against the alternative $\mu = 25$.)

Chapter 7 Solutions

7.1. (a) The standard error of the mean is $\frac{s}{\sqrt{n}} = \frac{\$96}{\sqrt{16}} = \$24$. **(b)** The degrees of freedom are df $= n - 1 = 15$.

7.3. For the mean monthly rent, the 95% confidence interval for μ is

$$\$613 \pm 2.131 \left(\frac{\$96}{\sqrt{16}} \right) = \$613 \pm \$51.14 = \$561.86 \text{ to } \$664.14$$

7.5. (a) Yes, $t = 2.18$ is significant when $n = 18$. This can be determined either by comparing to the df $= 17$ line in Table D (where we see that $t > 2.110$, the 2.5% critical value) or by computing the two-sided P-value (which is $P = 0.0436$). **(b)** No, $t = 2.18$ is not significant when $n = 10$, as can be seen by comparing to the df $= 9$ line in Table D (where we see that $t < 2.262$, the 2.5% critical value) or by computing the two-sided P-value (which is $P = 0.0572$). **(c)** Student sketches will likely be indistinguishable from Normal distributions; careful students may try to show that the $t(9)$ distribution is shorter in the center and heavier to the left and right ("in the tails") than the $t(17)$ distribution (as is the case here), but in reality, the difference is nearly imperceptible.

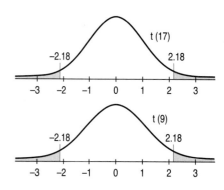

7.7. Software will typically give a more accurate value for t^* than that given in Table D, and will not round off intermediate values such as the standard deviation. Otherwise, the details of this computation are the same as what is shown in the textbook: df $= 7$, $t^* = 2.3646$, $6.75 \pm t^*(3.8822/\sqrt{8}) = 6.75 \pm 3.2456 = 3.5044$ to 9.9956, or about 3.5 to 10.0 hours per month.

7.9. About -1.33 to 10.13: Using the mean and standard deviation from the previous exercise, the 95% confidence interval for μ is

$$4.4 \pm 2.7765 \left(\frac{4.6152}{\sqrt{5}} \right) = 4.4 \pm 5.7305 = -1.3305 \text{ to } 10.1305$$

(This is the interval produced by software; using the critical value $t^* = 2.776$ from Table D gives -1.3296 to 10.1296.)

7.11. The distribution is clearly non-Normal, but the sample size ($n = 63$) should be sufficient to overcome this, especially in the absence of strong skewness. One might question the independence of the observations; it seems likely that after 40 or so tickets had been posted for sale, that someone listing a ticket would look at those already posted for an idea of what price to charge.

   If we were to use t procedures, we would presumably take the viewpoint that these 63 observations come from a larger population of hypothetical tickets for this game, and we are trying to estimate the mean μ of that population. However, because (based on the histogram in Figure 1.33) the population distribution is likely bimodal, the mean μ might not be the most useful summary of a bimodal distribution.

7.13. As was found in Example 7.9, we reject H_0 if $t = \frac{\bar{x}}{1.5/\sqrt{20}} \geq 1.729$, which is equivalent to $\bar{x} \geq 0.580$. The power we seek is $P(\bar{x} \geq 0.580$ when $\mu = 1.1)$, which is:

$$P\left(\frac{\bar{x} - 1.1}{1.5/\sqrt{20}} \geq \frac{0.580 - 1.1}{1.5/\sqrt{20}}\right) = P(Z \geq -1.55) = 0.9394$$

7.15. **(a)** df $= 10$, $t^* = 2.228$. **(b)** df $= 21$, $t^* = 2.080$. **(c)** df $= 21$, $t^* = 1.721$. **(d)** For a given confidence level, t^* (and therefore the margin of error) decreases with increasing sample size. For a given sample size, t^* increases with increasing confidence.

7.17. The 5% critical value for a t distribution with df $= 17$ is 1.740. Only one of the one-sided options (reject H_0 when $t > 1.740$) is shown; the other is simply the mirror image of this sketch (shade the area to the left of -1.740, and reject when $t < -1.740$).

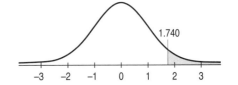

7.19. $\bar{x} = -15.3$ would support the alternative $\mu < 0$, and for that alternative, the P-value would still be 0.037. For the alternative $\mu > 0$ given in Exercise 7.18, the P-value is 0.963. Note that in the sketch shown, no scale has been given, because in the absence of a sample size, we do not know the degrees of freedom. Nevertheless, the P-value for the

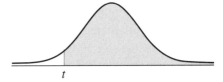

alternative $\mu > 0$ is the area above the computed value of the test statistic t, which will be the opposite of that found when $\bar{x} = 15.3$. As the area below t is 0.037, the area above this point must be 0.963.

7.21. **(a)** df $= 27$. **(b)** $1.703 < t < 2.052$. **(c)** Because the alternative is two-sided, we double the upper-tail probabilities to find the P-value: $0.05 < P < 0.10$. **(d)** $t = 2.01$ is not significant at either level (5% or 1%). **(e)** From software, $P \doteq 0.0546$.

7.23. Let P be the given (two-sided) P-value, and suppose that the alternative is $\mu > \mu_0$. If $\bar{x}$ is greater than μ_0, this supports the alternative over H_0. However, if $\bar{x} < \mu_0$, we would not take this as evidence against H_0 because $\bar{x}$ is on the "wrong" side of μ_0. So, if the value of $\bar{x}$ is on the "correct" side of μ_0, the one-sided P-value is simply $P/2$. However, if the value of $\bar{x}$ is on the "wrong" side of μ_0, the one-sided P-value is $1 - P/2$ (which will always be at least 0.5, so it will never indicate significant evidence against H_0).

7.25. **(a)** If μ is the mean number of uses a person can produce in 5 minutes after witnessing rudeness, we wish to test H_0: $\mu = 10$ versus H_a: $\mu < 10$. **(b)** $t = \frac{7.88 - 10}{2.35/\sqrt{34}} \doteq -5.2603$, with df $= 33$, for which $P < 0.0001$. This is very strong evidence that witnessing rudeness decreases performance.

7.27. (a) A stemplot (right) reveals that the distribution has two peaks and a high value (not quite an outlier). Both the stemplot and quantile plot show that the distribution is not Normal. The five-number summary is 2.2, 10.95, 28.5, 41.9, 69.3 (all in cm); a boxplot is not shown, but the long "whisker" between Q_3 and

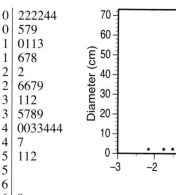

0	222244
0	579
1	0113
1	678
2	2
2	6679
3	112
3	5789
4	0033444
4	7
5	112
5	
6	
6	9

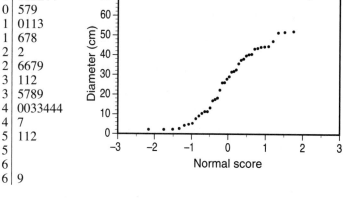

the maximum is an indication of the skewness. **(b)** Maybe: We have a large enough sample to overcome the non-Normal distribution, but we are sampling from a small population. **(c)** The mean is $\bar{x} = 27.29$ cm, $s \doteq 17.7058$ cm, and the margin of error is $t^* \cdot s/\sqrt{40}$:

	df	t^*	Interval
Table D	30	2.042	$27.29 \pm 5.7167 = 21.57$ to 33.01 cm
Software	39	2.0227	$27.29 \pm 5.6626 = 21.63$ to 32.95 cm

(d) One could argue for either answer. We chose a random sample from this tract, so the main question is, can we view trees in this tract as being representative of trees elsewhere?

7.29. (a) The distribution is not Normal—there were lots of 1s and 10s—but the nature of the scale means that there can be no extreme outliers, so with a sample of size 60, the t methods should be acceptable. **(b)** The mean is $\bar{x} \doteq 5.9$, $s \doteq 3.7719$, and the margin of error is $t^* \cdot s/\sqrt{60}$:

1	0000000000000000
2	0000
3	0
4	0
5	00000
6	000
7	0
8	000000
9	00000
10	00000000000000000000

	df	t^*	Interval
Table D	50	2.009	$5.9 \pm 0.9783 = 4.9217$ to 6.8783
Software	59	2.0010	$5.9 \pm 0.9744 = 4.9256$ to 6.8744

(c) Because this is not a random sample, it may not represent other children well.

7.31. These intervals are computed as $\bar{x} \pm t^* \cdot s/\sqrt{282}$. We see that the width of the interval increases with confidence level.

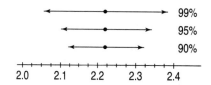

		df	t^*	Interval
90% confidence	Table D	100	1.660	$2.22 \pm 0.1018 = 2.1182$ to 2.3218
	Software	281	1.6503	$2.22 \pm 0.1012 = 2.1188$ to 2.3212
95% confidence	Table D	100	1.984	$2.22 \pm 0.1217 = 2.0983$ to 2.3417
	Software	281	1.9684	$2.22 \pm 0.1207 = 2.0993$ to 2.3407

7.33. (a) $t = \frac{328-0}{256/\sqrt{16}} \doteq 5.1250$ with df $= 15$, for which $P \doteq 0.0012$. There is strong evidence of a change in NEAT. **(b)** With $t^* = 2.131$, the 95% confidence interval is 191.6 to 464.4 kcal/day. This tells us how much of the additional calories might have been burned by the increase in NEAT: It consumed 19% to 46% of the extra 1000 kcal/day.

7.35. (a) We wish to test $H_0\colon \mu_c = \mu_d$ versus $H_a\colon \mu_c \neq \mu_d$, where μ_c is the mean computer-calculated mpg and μ_d is the mean mpg computed by the driver. Equivalently, we can state the hypotheses in terms of μ, the mean difference between computer- and driver-calculated mpgs, testing $H_0\colon \mu = 0$ versus $H_a\colon \mu \neq 0$. **(b)** With mean difference $\bar{x} \doteq 2.73$ and standard deviation $s \doteq 2.8015$, the test statistic is $t = \frac{2.73-0}{2.8015/\sqrt{20}} \doteq 4.3580$ with df $= 19$, for which $P \doteq 0.0003$. We have strong evidence that the results of the two computations are different.

7.37. (a) To test $H_0\colon \mu = 925$ picks versus $H_a\colon \mu > 925$ picks, we have $t = \frac{938.\overline{2}-925}{24.2971/\sqrt{36}} \doteq 3.27$ with df $= 35$, for which $P \doteq 0.0012$. **(b)** For $H_0\colon \mu = 935$ picks versus $H_a\colon \mu > 935$ picks, we have $t = \frac{938.\overline{2}-935}{24.2971/\sqrt{36}} \doteq 0.80$, again with df $= 35$, for which $P \doteq 0.2158$. **(c)** The 90% confidence interval from the previous exercise was 931.4 to 945.1 picks, which includes 935, but not 925. For a test of $H_0\colon \mu = \mu_0$ versus $H_a\colon \mu \neq \mu_0$, we know that $P < 0.10$ for values of μ_0 outside the interval, and $P > 0.10$ if μ_0 is inside the interval. The one-sided P-value would be half of the two-sided P-value.

7.39. (a) The differences are spread from -0.018 to 0.020 g, with mean $\bar{x} = -0.0015$ and standard deviation $s \doteq 0.0122$ g. A stemplot is shown on the right; the sample is too small to make judgments about skewness or symmetry. **(b)** For $H_0\colon \mu = 0$ versus $H_a\colon \mu \neq 0$, we find $t = \frac{-0.0015-0}{s/\sqrt{8}} \doteq -0.347$ with df $= 7$, for which $P = 0.7388$. We cannot reject H_0 based on this sample. **(c)** The 95% confidence interval for μ is

```
-1 | 85
-1 |
-0 | 65
-0 |
 0 | 2
 0 | 55
 1 |
 1 |
 2 | 0
```

$$-0.0015 \pm 2.365 \left(\frac{0.0122}{\sqrt{8}} \right) = -0.0015 \pm 0.0102 = -0.0117 \text{ to } 0.0087 \text{ g}$$

(d) The subjects from this sample may be representative of future subjects, but the test results and confidence interval are suspect because this is not a random sample.

7.41. (a) We test $H_0\colon \mu = 0$ versus $H_a\colon \mu > 0$, where μ is the mean change in score (that is, the mean improvement). **(b)** The distribution is slightly left-skewed, with mean $\bar{x} = 2.5$ and $s \doteq 2.8928$. **(c)** $t = \frac{2.5-0}{s/\sqrt{20}} \doteq 3.8649$, df $= 19$, and $P = 0.0005$; there is strong evidence of improvement in listening test scores. **(d)** With df $= 19$, we have $t^* = 2.093$, so the 95% confidence interval is 1.1461 to 3.8539.

```
-0 | 6
-0 |
-0 |
-0 | 0
 0 | 0011
 0 | 222333333
 0 |
 0 | 66666
```

7.43. The distribution is fairly symmetrical with no outliers. The mean IQ is $\bar{x} = 114.98\overline{3}$ and the standard deviation is $s \doteq 14.8009$. The 95% confidence interval is $\bar{x} \pm t^*(s/\sqrt{60})$, which is about 111.1 to 118.8:

	df	t^*	Interval
Table D	50	2.009	111.1446 to 118.8221
Software	59	2.0010	111.1598 to 118.8068

Because all students in the sample came from the same school, this *might* adequately describe the mean IQ at this school, but the sample could not be considered representative of all fifth graders.

```
 8 | 12
 8 | 9
 9 | 04
 9 | 67
10 | 01112223
10 | 568999
11 | 0002233444
11 | 5677788
12 | 223444
12 | 56778
13 | 01344
13 | 6799
14 | 2
14 | 5
```

7.45. We test H_0: median $= 0$ versus H_a: median > 0—or equivalently, H_0: $p = 1/2$ versus H_a: $p > 1/2$, where p is the probability that Jocko's estimate is higher. One difference is 0; of the nine non-zero differences, seven are positive. The P-value is $P(X \geq 7) = 0.0898$ from a $B(9, 0.5)$ distribution; there is not quite enough evidence to conclude that Jocko's estimates are higher. In Exercise 7.34 we were able to reject H_0; here we cannot.

Note: *The failure to reject H_0 in this case is because with the sign test, we pay attention only to the* sign *of each difference, not the* size. *In particular, the negative differences are each given the same "weight" as each positive difference, in spite of the fact that the negative differences are only $-\$50$ and $-\$75$, while most of the positive differences are larger. See the "Caution" about the sign test on page 425 of the text.*

Minitab output: Sign test of median = 0 versus median > 0
```
         N  BELOW  EQUAL  ABOVE  P-VALUE   MEDIAN
Diff    10    2      1      7    0.0898    125.0
```

7.47. We test H_0: median $= 0$ versus H_a: median $\neq 0$. There were three negative and five positive differences, so the P-value is $2P(X \geq 5)$ for a binomial distribution with parameters $n = 8$ and $p = 0.5$. From Table C or software (Minitab output below), we have $P = 0.7266$, which gives no reason to doubt H_0. The t test P-value was 0.6410.

Minitab output: Sign test of median = 0 versus median ≠ 0
```
          N  BELOW  EQUAL  ABOVE  P-VALUE   MEDIAN
opdiff    8    3      0      5    0.7266    3.500
```

7.49. We test H_0: median $= 0$ versus H_a: median > 0, or H_0: $p = 1/2$ versus H_a: $p > 1/2$. Out of the 20 differences, 17 are positive (and none equal 0). The P-value is $P(X \geq 17)$ for a $B(20, 0.5)$ distribution. From Table C or software (Minitab output below), we have $P = 0.0013$, so we reject H_0 and conclude that the results of the two computations are different. (Using a t test, we found $P \doteq 0.0003$, which led to the same conclusion.)

Minitab output: Sign test of median = 0 versus median > 0
```
        N  BELOW  EQUAL  ABOVE  P-VALUE   MEDIAN
diff   20    3      0     17    0.0013    3.000
```

7.51. The standard deviation for the given data was $s \doteq 0.012224$. With $\alpha = 0.05$, $t = \frac{\bar{x}}{s/\sqrt{15}}$, and df $= 14$, we reject H_0 if $|t| \geq 2.145$, which means $|\bar{x}| \geq (2.145)(s/\sqrt{15})$, or

$|\bar{x}| \geq 0.00677$. Assuming $\mu = 0.002$:

$$P(|\bar{x}| \geq 0.00677) = 1 - P(-0.00677 \leq \bar{x} \leq 0.00677)$$

$$= 1 - P\left(\frac{-0.00677 - 0.002}{s/\sqrt{15}} \leq \frac{\bar{x} - 0.002}{s/\sqrt{15}} \leq \frac{0.00677 - 0.002}{s/\sqrt{15}}\right)$$

$$= 1 - P(-2.78 \leq Z \leq 1.51)$$

$$= 1 - (0.9345 - 0.0027) \doteq 0.07$$

The power is about 7% against this alternative—not surprising, given the small sample size, and the fact that the difference (0.002) is small relative to the standard deviation.

   Note: *Power calculations are often done with software. This may give answers that differ slightly from those found by the method described in the text. Most software does these computations with a "noncentral t distribution" (used in the text for two-sample power problems) rather than a Normal distribution, resulting in more accurate answers. In most situations, the practical conclusions drawn from the power computations are the same regardless of the method used.*

7.53. Taking $s = 1.5$ as in Example 7.9, the power for the alternative $\mu = 0.75$ is:

$$P\left(\bar{x} \geq \frac{t^*s}{\sqrt{n}} \text{ when } \mu = 0.75\right) = P\left(\frac{\bar{x} - 0.75}{s/\sqrt{n}} \geq \frac{t^*s/\sqrt{n} - 0.75}{s/\sqrt{n}}\right) = P\left(Z \geq t^* - 0.5\sqrt{n}\right)$$

Using trial-and-error, we find that with $n = 26$, power $\doteq 0.7999$, and with $n = 27$, power $\doteq 0.8139$. Therefore, we need $n > 26$.

7.55. Because $2.264 < t < 2.624$ and the alternative is two-sided, Table D tells us that the P-value is $0.02 < P < 0.04$. (Software gives $P = 0.0280$.) That is sufficient to reject H_0 at $\alpha = 0.05$.

7.57. We find $\text{SE}_D \doteq 4.5607$. The options for the 95% confidence interval for $\mu_1 - \mu_2$ are shown on the right. The instructions for this exercise say to use the second approximation (df $= 9$), in which

df	t^*	Confidence interval
15.7	2.1236	-19.6851 to -0.3149
9	2.262	-20.3163 to 0.3163

case we do not reject H_0, because 0 falls in the the 95% confidence interval. Using the first approximation (df $= 15.7$, typically given by software), the interval is narrower, and we would reject H_0 at $\alpha = 0.05$ against a two-sided alternative. (In fact, $t \doteq -2.193$, for which $0.05 < P < 0.1$ [Table D, df $= 9$], or $P \doteq 0.0438$ [software, df $\doteq 15.7$].)

7.59. SPSS and SAS give both results (the SAS output refers to the unpooled result as the Satterthwaite method), while JMP and Excel show only the unpooled procedures. The pooled t statistic is 1.998, for which $P = 0.0808$.

   Note: *When the sample sizes are equal—as in this case—the pooled and unpooled t statistics are equal. (See the next exercise.)*

   Both Excel and JMP refer to the unpooled test with the slightly-misleading phrase "assuming unequal variances." The SAS output also implies that the variances are unequal for this method. In fact, unpooled procedures make no *assumptions about the variances.*

   Finally, note that both Excel and JMP can do pooled procedures as well as the unpooled procedures that are shown.

7.61. (a) Hypotheses should involve μ_1 and μ_2 (population means) rather than $\bar{x}_1$ and $\bar{x}_2$ (sample means). **(b)** The samples are not independent; we would need to compare the 56 males to the 44 females. **(c)** We need P to be small (for example, less than 0.10) to reject H_0. A large P-value like this gives no reason to doubt H_0. **(d)** Assuming the researcher computed the t statistic using $\bar{x}_1 - \bar{x}_2$, a positive value of t does not support H_a. (The one-sided P-value would be 0.982, not 0.018.)

7.63. (a) We cannot reject H_0: $\mu_1 = \mu_2$ in favor of the two-sided alternative at the 5% level because $0.05 < P < 0.10$ (Table D) or $P \doteq 0.0542$ (software). **(b)** We could reject H_0 in favor of H_a: $\mu_1 < \mu_2$. A negative t-statistic means that $\bar{x}_1 < \bar{x}_2$, which supports the claim that $\mu_1 < \mu_2$, and the one-sided P-value would be half of its value from part (a): $0.025 < P < 0.05$ (Table D) or $P \doteq 0.0271$ (software).

7.65. (a) Stemplots (right) do not look particularly Normal, but they have no extreme outliers or skewness, so t procedures should be reasonably safe. **(b)** The table of summary statistics is below on the left. **(c)** We wish to test H_0: $\mu_N = \mu_S$ versus H_a: $\mu_N < \mu_S$. **(d)** We find $SE_D \doteq 0.3593$ and $t = -4.303$, so $P \doteq 0.0001$ (df $\doteq 26.5$) or $P < 0.0005$ (df $= 13$). Either way, we reject H_0. **(e)** The 95% confidence interval for the difference is one of the two options in the table below on the right.

	Neutral		Sad
0	0000000	0	0
0	55	0	5
1	000	1	000
1		1	555
2	00	2	0
		2	55
		3	00
		3	55
		4	00

Group	n	$\bar{x}$	s		df	t^*	Confidence interval
Neutral	14	$0.5714	$0.7300		26.5	2.0538	−2.2842 to −0.8082
Sad	17	$2.1176	$1.2441		13	2.160	−2.3224 to −0.7701

7.67. (a) The female means and standard deviations are $\bar{x}_F \doteq 4.0791$ and $s_F \doteq 0.9861$; for males, they are $\bar{x}_M \doteq 3.8326$ and $s_M \doteq 1.0677$. **(b)** Both distributions are somewhat skewed to the left. This

df	t^*	Confidence interval
402.2	1.9659	0.0793 to 0.4137
220	1.9708	0.0788 to 0.4141
100	1.984	0.0777 to 0.4153

can be seen by constructing a histogram, but is also evident in the data table in the text by noting the large numbers of "4" and "5" ratings for both genders. However, because the ratings range from 1 to 5, there are no outliers, so the t procedures should be safe. **(c)** We find $SE_D \doteq 0.0851$ and $t \doteq 2.898$, for which $P \doteq 0.0040$ (df $\doteq 402.2$) or $0.002 < P < 0.005$ (df $= 220$). Either way, there is strong evidence of a difference in satisfaction. **(d)** The 95% confidence interval for the difference is one of the three options in the table on the right—roughly 0.08 to 0.41. **(e)** While we have evidence of a difference in mean ratings, it might not be as large as 0.25.

7.69. (a) Assuming we have SRSs from each population, use of two-sample t procedures seems reasonable. (We cannot assess Normality, but the large sample sizes would overcome most problems.)

df	t^*	Confidence interval
76.1	1.9916	−11.7508 to 15.5308
36	2.0281	−12.0005 to 15.7805
30	2.042	−12.0957 to 15.8757

(b) We wish to test H_0: $\mu_f = \mu_m$ versus H_a: $\mu_f \neq \mu_m$. **(c)** We find SE$_D \doteq 6.8490$ mg/dl. The test statistic is $t \doteq 0.276$, with df $\doteq 76.1$ (or 36—use 30 for Table D), for which $P \doteq 0.78$. We have no reason to believe that male and female cholesterol levels are different. **(d)** The options for the 95% confidence interval for $\mu_f - \mu_m$ are shown on the right. **(e)** It might not be appropriate to treat these students as SRSs from larger populations.

   Note: *Because t distributions are more spread out than Normal distributions, a t-value that would not be significant for a Normal distribution (such as 0.276) cannot possibly be significant when compared to a t distribution.*

7.71. (a) The distribution cannot be Normal because all numbers are integers. **(b)** The t procedures should be appropriate because we have two large samples with no outliers. **(c)** We will test H_0: $\mu_I = \mu_C$

df	t^*	Confidence interval
354.0	1.9667	0.5143 to 0.9857
164	1.9745	0.5134 to 0.9866
100	1.984	0.5122 to 0.9878

versus H_a: $\mu_I > \mu_C$ (or $\mu_I \neq \mu_C$). The one-sided alternative reflects the researchers' (presumed) belief that the intervention would increase scores on the test. The two-sided alternative allows for the possibility that the intervention might have had a negative effect. **(d)** SE$_D = \sqrt{s_I^2/n_I + s_C^2/n_C} \doteq 0.1198$ and $t = (\bar{x}_I - \bar{x}_C)/\text{SE}_D \doteq 6.258$. Regardless of how we compute degrees of freedom (df $\doteq 354$ or 164), the P-value is very small: $P < 0.0001$. We reject H_0 and conclude that the intervention increased test scores. **(e)** The interval is $\bar{x}_I - \bar{x}_C \pm t^*\text{SE}_D$; the value of t^* depends on the df (see the table), but note that in every case the interval rounds to 0.51 to 0.99. **(f)** The results for this sample may not generalize well to other areas of the country.

7.73. (a) This may be near enough to an SRS, if this company's working conditions were similar to that of other workers. **(b)** SE$_D \doteq 0.7626$; regardless of how we choose df, the interval rounds to 9.99 to

df	t^*	Confidence interval
137.1	1.9774	9.9920 to 13.0080
114	1.9810	9.9893 to 13.0107
100	1.984	9.9870 to 13.0130

13.01 mg.y/m^3. **(c)** A one-sided alternative would seem to be reasonable here; specifically, we would likely expect that the mean exposure for outdoor workers would be lower. For testing H_0, we find $t = 15.08$, for which $P < 0.0001$ with either df $= 137$ or 114 (and for either a one- or a two-sided alternative). We have strong evidence that outdoor concrete workers have lower dust exposure than the indoor workers. **(d)** The sample sizes are large enough that skewness should not matter.

7.75. To find a confidence interval $(\bar{x}_1 - \bar{x}_2) \pm t^*\text{SE}_D$, we need one of the following:

- Sample sizes and standard deviations—in which case we could find the interval in the usual way
- t and df—because $t = (\bar{x}_1 - \bar{x}_2)/\text{SE}_D$, so we could compute SE$_D = (\bar{x}_1 - \bar{x}_2)/t$ and use df to find t^*
- df and a more accurate P-value—from which we could determine t, and then proceed as above

The confidence interval could give us useful information about the magnitude of the difference (although with such a small *P*-value, we do know that a 95% confidence interval would not include 0).

7.77. This is a matched pairs design; for example, Monday hits are (at least potentially) not independent of one another. The correct approach would be to use one-sample *t* methods on the seven differences (Monday hits for design 1 minus Monday hits for design 2, Tuesday/1 minus Tuesday/2, and so on).

7.79. The next 10 employees who need screens might not be an independent group—perhaps they all come from the same department, for example. Randomization reduces the chance that we end up with such unwanted groupings.

7.81. (a) Stemplots, boxplots, and five-number summaries (in cm) are shown on the right. The north distribution is right-skewed, while the south distribution is left-skewed. (b) The methods of this section seem to be appropriate in spite of the skewness because the sample sizes are relatively large, and there are no outliers in either distribution. (c) We test H_0: $\mu_n = \mu_s$ versus H_a: $\mu_n \neq \mu_s$; we should use a two-sided alternative because we have no reason (before looking at the data) to expect a difference in a particular direction. (d) The means and standard deviations are $\bar{x}_n = 23.7$, $s_n \doteq 17.5001$, $\bar{x}_s = 34.5\overline{3}$, and $s_s \doteq 14.2583$ cm. Then $\mathrm{SE}_D \doteq 4.1213$, so $t = -2.629$ with df $= 55.7$ ($P = 0.011$) or df $= 29$ ($P = 0.014$). We conclude that the means are different (specifically, the south mean is greater than the north mean). (e) See the table for possible 95% confidence intervals.

North		South
43322	0	2
65	0	57
443310	1	2
955	1	8
	2	13
8755	2	689
0	3	2
996	3	566789
43	4	003444
6	4	578
4	5	0112
85	5	

Tree diameter (cm)

	Min	Q_1	M	Q_3	Max
North	2.2	10.2	17.05	39.1	58.8
South	2.6	26.1	37.70	44.6	52.9

df	t^*	Confidence interval
55.7	2.0035	−19.0902 to −2.5765
29	2.045	−19.2614 to −2.4053

7.83. (a) $\mathrm{SE}_D \doteq 1.9686$. Answers will vary with the df used (see the table), but the interval is roughly −1 to 7 units. (b) Because of random fluctuations between stores, we might (just by chance) have seen a rise in the average number of units sold even if actual mean sales had remained unchanged. (Based on the confidence interval, mean sales might have even dropped slightly.)

df	t^*	Confidence interval
122.5	1.9795	−0.8968 to 6.8968
54	2.0049	−0.9468 to 6.9468
50	2.009	−0.9549 to 6.9549

7.85. (a) We test $H_0: \mu_b = \mu_f$; $H_a: \mu_b > \mu_f$.
$SE_D \doteq 0.5442$ and $t = 1.654$, for which $P = 0.0532$
(df $= 37.6$) or 0.0577 (df $= 18$); there is not quite

df	t^*	Confidence interval
37.6	2.0251	-0.2021 to 2.0021
18	2.101	-0.2434 to 2.0434

enough evidence to reject H_0 at $\alpha = 0.05$. **(b)** The confidence interval depends on the degrees of freedom used; see the table. **(c)** We need two independent SRSs from Normal populations.

7.87. A table of means and standard deviations is on the right. The pooled standard deviation is $s_p \doteq 0.9347$, so the pooled standard error is $s_p\sqrt{1/22 + 1/20} \doteq 0.2888$.

Group	n	$\bar{x}$	s
Primed	22	4.00	0.9258
Non-primed	20	2.95	0.9445

The test statistic is $t \doteq 3.636$ with df $= 40$, for which $P = 0.0008$, and the 95% confidence interval (with $t^* \doteq 2.021$) is 0.4663 to 1.6337.

In the solution to Exercise 7.66, we reached the same conclusion on the significance test ($t \doteq 3.632$ and $P \doteq 0.0008$) and the confidence interval (using the more-accurate df $\doteq 39.5$) was quite similar: 0.4655 to 1.6345.

7.89. See the solution to Exercise 7.81 for means and standard deviations. The pooled standard deviation is $s_p \doteq 15.9617$, and the standard error is

df	t^*	Confidence interval
58	2.0017	-19.0830 to -2.5837
50	2.009	-19.1130 to -2.5536

$SE_D \doteq 4.1213$. For the significance test, $t = -2.629$, df $= 58$, and $P = 0.0110$, so we have fairly strong evidence (though not quite significant at $\alpha = 0.01$) that the south mean is greater than the north mean. Possible answers for the confidence interval (with software, and with Table D) are given in the table. All results are similar to those found in Exercise 7.81.

Note: *If $n_1 = n_2$ (as in this case), the standard error and t statistic are the same for the usual and pooled procedures. The degrees of freedom will usually be different (specifically, df is larger for the pooled procedure, unless $s_1 = s_2$ and $n_1 = n_2$).*

7.91. With $s_n \doteq 17.5001$, $s_s \doteq 14.2583$, and $n_n = n_s = 30$, we have $s_n^2/n_n \doteq 10.2085$ and $s_s^2/n_s \doteq 6.7767$, so:

$$\text{df} = \frac{\left(\frac{s_n^2}{n_n} + \frac{s_s^2}{n_s}\right)^2}{\frac{1}{n_n-1}\left(\frac{s_n^2}{n_n}\right)^2 + \frac{1}{n_s-1}\left(\frac{s_s^2}{n_s}\right)^2} \doteq \frac{(10.2085 + 6.7767)^2}{\frac{1}{29}(10.2085^2 + 6.7767^2)} \doteq 55.7251$$

7.93. (a) With $s_i \doteq 7.8$, $n_i = 115$, $s_o \doteq 3.4$, and $n_o = 220$, we have $s_i^2/n_i \doteq 0.5290$ and $s_o^2/n_o \doteq 0.05455$, so:

$$\mathrm{df} \doteq \frac{(0.5290 + 0.05455)^2}{\dfrac{0.5290^2}{114} + \dfrac{0.05455^2}{219}} \doteq 137.0661$$

	df	t^*	Confidence interval
Part (d)	333	1.9671	10.2931 to 12.7069
	100	1.984	10.2827 to 12.7173
Part (e)	333	1.9671	4.5075 to 5.2925
	100	1.984	4.5042 to 5.2958

(b) $s_p = \sqrt{\dfrac{(n_i-1)s_i^2 + (n_o-1)s_o^2}{n_i + n_o - 2}} \doteq 5.3320$, which is slightly closer to s_o (the standard deviation from the larger sample). **(c)** With no assumption of equality, $\mathrm{SE}_1 = \sqrt{s_i^2/n_i + s_o^2/n_o} \doteq 0.7626$. With the pooled method, $\mathrm{SE}_2 = s_p\sqrt{1/n_i + 1/n_o} \doteq 0.6136$. **(d)** With the pooled standard deviation, $t \doteq 18.74$ and df = 333, for which $P < 0.0001$, and the 95% confidence interval is as shown in the table. With the smaller standard error, the t value is larger (it had been 15.08), and the confidence interval is narrower. The P-value is also smaller (although both are less than 0.0001). **(e)** With $s_i \doteq 2.8$, $n_i = 115$, $s_o \doteq 0.7$, and $n_o = 220$, we have $s_i^2/n_i \doteq 0.06817$ and $s_o^2/n_o \doteq 0.002227$, so:

$$\mathrm{df} \doteq \frac{(0.06817 + 0.002227)^2}{\dfrac{0.06817^2}{114} + \dfrac{0.002227^2}{219}} \doteq 121.5030$$

The pooled standard deviation is $s_p \doteq 1.7338$; the standard errors are $\mathrm{SE}_1 = 0.2653$ (with no assumptions) and $\mathrm{SE}_2 = 0.1995$ (assuming equal standard deviations). The pooled t is 24.56 (df = 333, $P < 0.0001$), and the 95% confidence intervals are shown in the table. The pooled and usual t procedures compare similarly to the results for part (d): With the pooled procedure, t is larger, and the interval is narrower.

7.95. (a) From an $F(15, 22)$ distribution with $\alpha = 0.05$, $F^* = 2.15$. **(b)** Because $F = 2.45$ is greater than the 5% critical value, but less than the 2.5% critical value ($F^* = 2.50$), we know that P is between $2(0.025) = 0.05$ and $2(0.05) = 0.10$. (Software tells us that $P = 0.055$.) $F = 2.45$ is significant at the 10% level but not at the 5% level.

7.97. The power would be smaller. A larger value of σ means that large differences between the sample means would arise more often by chance so that, if we observe such a difference, it gives less evidence of a difference in the population means.

    Note: *The table on the right shows the decrease in the power as σ increases.*

σ	Power
7.4	0.7965
7.5	0.7844
7.6	0.7722
7.7	0.7601
7.8	0.7477

7.99. The test statistic is $F = \left(\dfrac{9.9}{7.5}\right)^2 \doteq 1.742$, with df 60 and 176. The two-sided P-value is 0.0057, so we can reject H_0 and conclude that the standard deviations are different. We do not know if the distributions are Normal, so this test may not be reliable.

7.101. The test statistic is $F = \dfrac{1.16^2}{1.15^2} \doteq 1.0175$ with df 211 and 164. Table E tells us that $P > 0.20$, while software gives $P = 0.9114$. The distributions are not Normal ("total score was an integer between 0 and 6"), so the test may not be reliable (although with s_1 and s_2 so close, the conclusion is probably correct). To reject at the 5% level, we would need $F > F^*$, where $F^* = 1.46$ (using df 120 and 100 from Table E) or $F^* = 1.3392$ (using software). As $F = s_2^2/s_1^2$, we would need $s_2^2 > s_1^2 F^*$, or $s_2 > 1.15\sqrt{F^*}$, which is about 1.3896 (Table E) or 1.3308 (software).

7.103. The test statistic is $F = \frac{7.8^2}{3.7^2} \doteq 5.2630$ with df 114 and 219. Table E tells us that $P < 0.002$, while software gives $P < 0.0001$; we have strong evidence that the standard deviations differ. The authors described the distributions as somewhat skewed, so the Normality assumption may be violated.

7.105. The test statistic is $F \doteq \frac{17.5001^2}{14.2583^2} \doteq 1.5064$ with df 29 and 29. Table E tells us that $P > 0.2$, while software gives $P - 0.2757$; we cannot conclude that the standard deviations differ. The stemplots and boxplots of the north/south distributions in Exercise 7.81 do not appear to be Normal (both distributions were skewed), so the results may not be reliable.

7.107. (a) To test H_0: $\sigma_1 = \sigma_2$ versus H_a: $\sigma_1 \neq \sigma_2$, we find $F \doteq \frac{7.1554^2}{6.7676^2} \doteq$ 1.1179. We do not reject H_0. **(b)** With an $F(4, 4)$ distribution with a two-sided alternative, we need the critical value for $p = 0.025$: $F^* = 9.60$. The table on the right gives the critical values for other sample sizes. With such small samples, this is a very low-power test; large differences between σ_1 and σ_2 would rarely be detected.

n	F^*
5	9.60
4	15.44
3	39.00
2	647.79

7.109. The four standard deviations from Exercises 7.81 and 7.82 are $s_n \doteq 17.5001$, $s_s \doteq 14.2583$, $s_e \doteq 16.0743$, and $s_w \doteq 15.3314$ cm. Using a larger σ for planning the study is advisable because it provides a conservative (safe) estimate of the power. For example, if we choose a sample size to provide 80% power and the true σ is smaller than that used for planning, the actual power of the test is greater than the desired 80%.

	Power with $n =$	
σ	20	60
15	0.5334	0.9527
16	0.4809	0.9255
17	0.4348	0.8928
18	0.3945	0.8560

Results of additional power computations depend on what students consider to be "other reasonable values of σ." Shown in the table are some possible answers using the Normal approximation. (Powers computed using the noncentral t distribution are slightly greater.)

7.111. The mean is $\bar{x} = 140.5$, the standard deviation is $s \doteq 13.58$, and the standard error of the mean is $s_{\bar{x}} \doteq 6.79$. It would not be appropriate to construct a confidence interval because we cannot consider these four scores to be an SRS.

7.113. The plot shows that t^* approaches 1.96 as df increases.

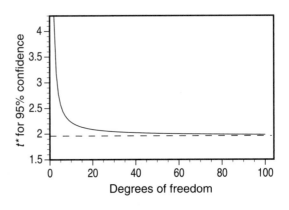

7.115. The margin of error is $t^*/\sqrt{n}$, using t^* for df $= n - 1$ and 95% confidence. For example, when $n = 5$, the margin of error is 1.2417, and when $n = 10$, it is 0.7154, and for $n = 100$, it is 0.1984. As we see in the plot, as sample size increases, margin of error decreases (toward 0, although it gets there very slowly).

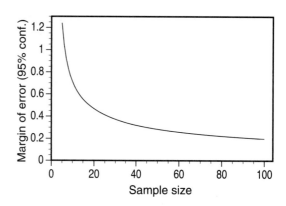

7.117. (a) Use two independent samples (students that live in the dorms, and those that live elsewhere). **(b)** Use a matched pairs design: Take a sample of college students, and have each subject rate the appeal of each label design. **(c)** Take a single sample of college students, and ask them to rate the appeal of the product.

7.119. (a) To test H_0: $\mu = 1.5$ versus H_a: $\mu < 1.5$, we have $t = \frac{1.20 - 1.5}{1.81/\sqrt{200}} \doteq -2.344$ with df $= 199$, for which $P \doteq 0.0100$. We can reject H_0 at the 5% significance level. **(b)** From Table D, use df $= 100$ and $t^* = 1.984$, so the 95% confidence interval for μ is

$$1.20 \pm 1.984 \left(\frac{1.81}{\sqrt{200}} \right) = 1.20 \pm 0.2539 = 0.9461 \text{ to } 1.4539 \text{ violations}$$

(With software, the interval is 0.9476 to 1.4524.) **(c)** While the significance test lets us conclude that there were fewer than 1.5 violations (on the average), the confidence interval gives us a range of values for the mean number of violations. **(d)** We have a large sample ($n = 200$), and the limited range means that there are no extreme outliers, so t procedures should be safe.

7.121. (a) The mean difference in body weight change (with wine minus without wine) was $\bar{x}_1 = 0.4 - 1.1 = -0.7$ kg, with standard error $SE_1 = 8.6/\sqrt{14} \doteq 2.2984$ kg. The mean difference in caloric intake was $\bar{x}_2 = 2589 - 2575 = 14$ cal, with $SE_2 = 210/\sqrt{14} \doteq 56.1249$ cal. **(b)** The t statistics $t_i = \bar{x}_i/SE_i$, both with df $= 13$, are $t_1 = -0.3046$ ($P_1 = 0.7655$) and $t_2 = 0.2494$ ($P_2 = 0.8069$). **(c)** For df $= 13$, $t^* = 2.160$, so the 95% confidence intervals $\bar{x}_i \pm t^*SE_i$ are -5.6646 to 4.2646 kg (-5.6655 to 4.2655 with software) and -107.2297 to 135.2297 cal (-107.2504 to 135.2504 with software). **(d)** Students might note a number of factors in their discussions; for example, all subjects were males, weighing 68 to 91 kg (about 150 to 200 lb), which may limit how widely we can extend these conclusions.

7.123. How much a person eats or drinks may depend on how many people he or she is sitting with. This means that the individual customers within each wine-label group probably cannot be considered to be independent of one another, which is a fundamental assumption of the t procedures.

7.125. The tables below contain summary statistics and 95% confidence intervals for the differences. For north/south differences, the test of H_0: $\mu_n = \mu_s$ gives $t = -7.15$ with df $= 575.4$ or 283; either way, $P < 0.0001$, so we reject H_0. For east/west differences, $t = -3.69$ with df $= 472.7$ or 230; either way, $P \doteq 0.0003$, so we reject H_0. The larger data set results in smaller standard errors (both are near 1.5, compared to about 4 in Exercises 7.81 and 7.82), meaning that t is larger and the margin of error is smaller.

	$\bar{x}$	s	n
North	21.7990	18.9230	300
South	32.1725	16.0763	284
East	24.5785	17.7315	353
West	30.3052	18.7264	231

	df	t^*	Confidence interval
N–S	575.4	1.9641	-13.2222 to -7.5248
	283	1.9684	-13.2285 to -7.5186
	100	1.984	-13.2511 to -7.4960
E–W	472.7	1.9650	-8.7764 to -2.6770
	230	1.9703	-8.7847 to -2.6687
	100	1.984	-8.8059 to -2.6475

7.127. **(a)** This is a matched pairs design because at each of the 24 nests, the same mockingbird responded on each day. **(b)** The variance of the difference is approximately $s_1^2 + s_4^2 - 2\rho s_1 s_4 = 48.684$, so the standard deviation is 6.9774 m. **(c)** To test H_0: $\mu_1 = \mu_4$ versus H_a: $\mu_1 \neq \mu_4$, we have $t = \frac{15.1 - 6.1}{6.9774/\sqrt{24}} \doteq 6.319$ with df $= 23$, for which P is very small. **(d)** Assuming the correlation is the same ($\rho = 0.4$), the variance of the difference is approximately $s_1^2 + s_5^2 - 2\rho s_1 s_5 = 31.324$, so the standard deviation is 5.5968 m. To test H_0: $\mu_1 = \mu_5$ versus H_a: $\mu_1 \neq \mu_5$, we have $t = \frac{4.9 - 6.1}{5.5968/\sqrt{24}} \doteq -1.050$ with df $= 23$, for which $P \doteq 0.3045$. **(e)** The significant difference between day 1 and day 4 suggests that the mockingbirds altered their behavior when approached by the same person for four consecutive days; seemingly, the birds perceived an escalating threat. When approached by a new person on day 5, the response was not significantly different from day 1; this suggests that the birds saw the new person as less threatening than a return visit from the first person.

7.129. The mean and standard deviation of the 25 numbers are $\bar{x} = 78.32\%$ and $s \doteq 33.3563\%$, so the standard error is $SE_{\bar{x}} \doteq 6.6713\%$. For df $= 24$, Table D gives $t^* = 2.064$, so the 95% confidence interval is $\bar{x} \pm 13.7695\% = 64.5505\%$ to 92.0895% (with software, $t^* = 2.0639$ and the interval is $\bar{x} \pm 13.7688\% = 64.5512\%$ to 92.0888%). This seems to support the retailer's claim: The original supplier's price was higher between 65% to 93% of the time.

7.131. Back-to-back stemplots below. The distributions appear similar; the most striking difference is the relatively large number of boys with low GPAs. Testing the difference in GPAs (H_0: $\mu_b = \mu_g$; H_a: $\mu_b < \mu_g$), we obtain $SE_D \doteq 0.4582$ and $t = -0.91$, which is not significant, regardless of whether we use df = 74.9 ($P = 0.1839$) or 30 ($0.15 < P < 0.20$). The 95% confidence interval for the difference $\mu_b - \mu_g$ in GPAs is shown in the second table on the right.

		GPA		IQ	
	n	$\bar{x}$	s	$\bar{x}$	s
Boys	47	7.2816	2.3190	110.96	12.121
Girls	31	7.6966	1.7208	105.84	14.271

	df	t^*	Confidence interval
GPA	74.9	1.9922	-1.3277 to 0.4979
	30	2.042	-1.3505 to 0.5207
IQ	56.9	2.0025	-1.1167 to 11.3542
	30	2.042	-1.2397 to 11.4772

For the difference in IQs, we find $SE_D \doteq 3.1138$. With the same hypotheses as before, we find $t = 1.64$—fairly strong evidence, but not quite significant at the 5% level: $P = 0.0528$ (df = 56.9) or $0.05 < P < 0.10$ (df = 30). The 95% confidence interval for the difference $\mu_b - \mu_g$ in IQs is shown in the second table on the right.

```
GPA:      Girls │   │ Boys              IQ:      Girls │    │ Boys
                │ 0 │ 5                         42 │  7 │
                │ 1 │ 7                            │  7 │ 79
                │ 2 │ 4                            │  8 │
            4   │ 3 │ 689                       96 │  8 │ 03
            7   │ 4 │ 068                       31 │  9 │ 03
          952   │ 5 │ 0                         86 │  9 │ 77
         4200   │ 6 │ 019                   433320 │ 10 │ 0234
    988855432   │ 7 │ 1124556666899           875 │ 10 │ 556667779
       998731   │ 8 │ 001112238          44422211 │ 11 │ 00001123334
        95530   │ 9 │ 1113445567               98 │ 11 │ 556899
           17   │10 │ 57                        0 │ 12 │ 03344
                                                8 │ 12 │ 67788
                                               20 │ 13 │
                                                  │ 13 │ 6
```

7.133. It is reasonable to have a prior belief that people who evacuated their pets would score higher, so we test H_0: $\mu_1 = \mu_2$ versus H_a: $\mu_1 > \mu_2$. We find $SE_D \doteq 0.4630$ and $t = 3.65$, which gives

df	t^*	Confidence interval
237.0	1.9700	0.7779 to 2.6021
115	1.9808	0.7729 to 2.6071
100	1.984	0.7714 to 2.6086

$P < 0.0005$ no matter how we choose degrees of freedom (115 or 237.0). As one might suspect, people who evacuated their pets have a higher mean score.

One might also compute a 95% confidence interval for the difference; these are given in the table.

7.135. The similarity of the sample standard deviations suggests that the population standard deviations are likely to be similar. The pooled standard deviation is $s_p \doteq 436.368$ and $t \doteq -0.3533$, so $P = 0.3621$ (df = 179)—still not significant.

7.137. No: What we have is nothing like an SRS of the population of school corporations.

7.139. (a) We test H_0: $\mu_B = \mu_D$ versus H_a: $\mu_B < \mu_D$.

	n	$\bar{x}$	s
Basal	22	41.0455	5.6356
DRTA	22	46.7273	7.3884
Strat	22	44.2727	5.7668

Pooling might be appropriate for this problem, in which case $s_p \doteq 6.5707$. Whether or not we pool, $SE_D \doteq 1.9811$ and $t \doteq 2.87$ with df $= 42$ (pooled), 39.3, or 21, so $P = 0.0032$, or 0.0033, or 0.0046. We conclude that the mean score using DRTA is higher than the mean score with the Basal method. The difference in the average scores is 5.68; options for a 95% confidence interval for the difference $\mu_D - \mu_B$ are given in the table below. **(b)** We test H_0: $\mu_B = \mu_S$ versus H_a: $\mu_B < \mu_S$. If we pool, $s_p \doteq 5.7015$. Whether or not we pool, $SE_D \doteq 1.7191$ and $t \doteq 1.88$ with df $= 42$, 42.0, or 21, so $P = 0.0337$, or 0.0337, or 0.0372. We conclude that the mean score using Strat is higher than the Basal mean score. The difference in the average scores is 3.23; options for a 95% confidence interval for the difference $\mu_S - \mu_B$ are given in the table below.

df	t^*	Confidence interval for $\mu_D - \mu_B$	df	t^*	Confidence interval for $\mu_S - \mu_B$
39.3	2.0223	1.6754 to 9.6882	42.0	2.0181	−0.2420 to 6.6966
21	2.0796	1.5618 to 9.8018	21	2.0796	−0.3477 to 6.8023
21	2.080	1.5610 to 9.8026	21	2.080	−0.3484 to 6.8030
42	2.0181	1.6837 to 9.6799	42	2.0181	−0.2420 to 6.6965
40	2.021	1.6779 to 9.6857	40	2.021	−0.2470 to 6.7015

7.141. **(a)** The distributions can be compared using a back-to-back stemplot (shown), or two histograms, or side-by-side boxplots. Three-bedroom homes are right-skewed; four-bedroom homes are generally more expensive. The top two prices from the three-bedroom distribution qualify as outliers using the $1.5 \times IQR$ criterion. Boxplots are probably a poor choice for displaying the distributions because they leave out so much detail, but five-

3BR		4BR
99987	0	
4432211100	1	4
8655	1	678
1	2	024
976	2	68
	3	223
	3	9
	4	2

number summaries do illustrate that four-bedroom prices are higher at every level. Summary statistics (in units of $1000) are given in the table below. **(b)** For testing $H_0: \mu_3 = \mu_4$ versus $H_a: \mu_3 \neq \mu_4$, we have $t \doteq -4.475$ with either df $= 20.98$ ($P \doteq 0.0002$) or df $= 13$ ($P < 0.001$). We reject H_0 and conclude that the mean prices are different (specifically, that 4BR houses are more expensive). **(c)** The one-sided alternative $\mu_3 < \mu_4$ could have been justified because it would be reasonable to expect that four-bedroom homes would be more expensive. **(d)** The 95% confidence interval for the difference $\mu_4 - \mu_3$ is about $63,823 to $174,642 (df $= 20.97$) or $61,685 to $176,779 (df $= 13$). **(e)** While the data were not gathered from an SRS, it seems that they should be a fair representation of three- and four-bedroom houses in West Lafayette. (Even so, the small sample sizes, together with the skewness and the outliers in the three-bedroom data, should make us cautious about the t procedures. Additionally, we might question independence in these data: When setting the asking price for a home, sellers are almost certainly influenced by the asking prices for similar homes on the market in the area.)

	n	$\bar{x}$	s	Min	Q_1	M	Q_3	Max
3BR	23	147.561	61.741	79.5	100.0	129.9	164.9	295.0
4BR	14	266.793	87.275	149.9	189.0	259.9	320.0	429.9

Chapter 8 Solutions

8.1. (a) $n = 760$ banks. **(b)** $X = 283$ banks expected to acquire another bank.
(c) $\hat{p} = \frac{283}{760} \doteq 0.3724$.

8.3. (a) With $\hat{p} \doteq 0.3724$, $\text{SE}_{\hat{p}} = \sqrt{\hat{p}(1 - \hat{p})/760} \doteq 0.01754$. **(b)** The 95% confidence interval is $\hat{p} \pm 1.96\,\text{SE}_{\hat{p}} = 0.3724 \pm 0.0344$. **(c)** The interval is 33.8% to 40.7%.

8.5. For $z = 1.34$, the two-sided P-value is the area under a
standard Normal curve above 1.34 and below -1.34.

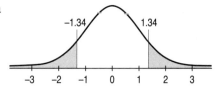

8.7. The sample proportion is $\hat{p} = \frac{15}{20} = 0.75$. To test H_0: $p = 0.5$ versus H_a: $p \neq 0.5$, the appropriate standard error is $\sigma_{\hat{p}} = \sqrt{p_0(1 - p_0)/20} \doteq 0.1118$, and the test statistic is $z = (\hat{p} - p_0)/\sigma_{\hat{p}} \doteq \frac{0.25}{0.1118} \doteq 2.24$. The two-sided P-value is 0.0250 (Table A) or 0.0253 (software), so this result is significant at the 5% level.

8.9. (a) To test H_0: $p = 0.5$ versus H_a: $p \neq 0.5$ with $\hat{p} = 0.35$, the test statistic is

$$z = \frac{\hat{p} - p_0}{\sqrt{\dfrac{p_0(1 - p_0)}{n}}} \doteq \frac{-0.15}{0.1118} \doteq -1.34$$

This is the opposite of the value of z given in Example 8.4, and the two-sided P-value is the same: 0.1802 (or 0.1797 with software). **(b)** The standard error for a confidence interval is $\text{SE}_{\hat{p}} = \sqrt{\hat{p}(1 - \hat{p})/20} \doteq 0.1067$, so the 95% confidence interval is $0.35 \pm 0.2090 = 0.1410$ to 0.5590. This is the complement of the interval shown in the Minitab output in Figure 8.2.

8.11. The plot is symmetric about 0.5, where it has its maximum. (The maximum margin of error always occurs at $\hat{p} = 0.5$, but the size of the maximum error depends on the sample size.)

    Note: *The first printing of the text asked students to plot sample proportion ($\hat{p}$) versus the margin of error (m), rather than m versus $\hat{p}$. Because $\hat{p}$ is the explanatory variable, the latter is more natural.*

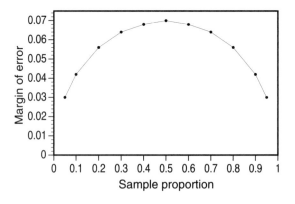

8.13. (a) Margin of error only accounts for random sampling error. **(b)** P-values measure the strength of the evidence against H_0, not the probability of it being true. **(c)** The confidence level cannot exceed 100%. (In practical terms, the confidence level must be *less than* 100%.)

8.15. The sample proportion is $\hat{p} = \frac{3274}{5000} \doteq 0.6548$, the standard error is $\text{SE}_{\hat{p}} \doteq 0.00672$, and the 95% confidence interval is $0.6548 \pm 0.0132 = 0.6416$ to 0.6680.

8.17. (a) $SE_{\hat{p}}$ depends on n, which is some number less than 7061. Without that number, we do not know the margin of error $z^*SE_{\hat{p}}$. **(b)** The number who expect to begin playing an instrument is 67% of half of 7061, or $(0.67)(0.5)(7061) \doteq 2365$ players. **(c)** Taking $n = (0.5)(7061) = 3530.5$, the 99% confidence interval is $\hat{p} \pm 2.576\,SE_{\hat{p}} = 0.67 \pm 0.02038 \doteq 0.6496$ to 0.6904. **(d)** We do not know the sampling methods used, which might make these methods unreliable.

    Note: *Even though n must be an integer in reality, it is not necessary to round n in part (c); the confidence interval formula works fine when n is not a whole number.*

8.19. If $\hat{p}$ has (approximately) a $N(p_0, \sigma)$ distribution under H_0, we reject H_0 (at the $\alpha = 0.05$ level) if $\hat{p}$ falls outside the range $p_0 \pm 1.96\sigma$. **(a)** If H_0 is true and $n = 60$, then $\sigma = \sqrt{(0.4)(0.6)/60} \doteq 0.06325$, so we reject H_0 when $\hat{p}$ is outside the range 0.2760 to 0.5240. Because $\hat{p} = \frac{x}{60}$, this corresponds to: Reject H_0 if $\hat{p} \le 0.2667$ ($x \le 15$) or $\hat{p} \ge 0.5333$ ($x \ge 32$). **(b)** If H_0 is true and $n = 100$, then $\sigma = \sqrt{(0.4)(0.6)/100} \doteq 0.04899$, so we reject H_0 when $\hat{p}$ is outside the range 0.3040 to 0.4960. This corresponds to: Reject H_0 if $\hat{p} \le 0.30$ ($x \le 30$) or $\hat{p} \ge 0.50$ ($x \ge 50$). **(c)** Shown on the right is one possible sketch, with the two Normal curves drawn on the same scale; the dashed curve and rejection cutoffs is for $n = 100$. (Most students will likely not realize that when σ is smaller, the curve must be taller to compensate for the decreased width.) With a larger sample size, smaller values of $|\hat{p} - 0.4|$ lead to the rejection of H_0.

8.21. (a) About $(0.42)(159{,}949) \doteq 67{,}179$ students plan to study abroad. **(b)** $SE_{\hat{p}} \doteq 0.00123$, the margin of error is $2.576\,SE_{\hat{p}} \doteq 0.00318$, and the 99% confidence interval is 0.4168 to 0.4232.

8.23. With $\hat{p} = 0.43$, we have $SE_{\hat{p}} \doteq 0.0131$, and the 95% confidence interval is $\hat{p} \pm 1.96\,SE_{\hat{p}} = 0.43 \pm 0.0257 \doteq 0.4043$ to 0.4557.

8.25. (a) $SE_{\hat{p}} = \sqrt{(0.87)(0.13)/430{,}000} \doteq 0.0005129$. For 99% confidence, the margin of error is $2.576\,SE_{\hat{p}} \doteq 0.001321$. **(b)** One source of error is indicated by the wide variation in response rates: We cannot assume that the statements of respondents represent the opinions of nonrespondents. The effect of the participation fee is harder to predict, but one possible impact is on the types of institutions that participate in the survey: Even though the fee is scaled for institution size, larger institutions can more easily absorb it. These other sources of error are much more significant than sampling error, which is the only error accounted for in the margin of error from part (a).

8.27. (a) The standard error is $SE_{\hat{p}} = \sqrt{(0.38)(0.62)/1048} \doteq 0.01499$, so the margin of error for 95% confidence is $1.96\,SE_{\hat{p}} \doteq 0.02939$ and the interval is 0.3506 to 0.4094. **(b)** Yes; some respondents might not admit to such behavior. The true frequency of such actions might be higher than this survey suggests.

8.29. (a) $\hat{p} = \frac{390}{1191} \doteq 0.3275$. The standard error is $SE_{\hat{p}} = \sqrt{\hat{p}(1 - \hat{p})/1191} \doteq 0.01360$, so the margin of error for 95% confidence is $1.96\,SE_{\hat{p}} \doteq 0.02665$ and the interval is 0.3008 to

0.3541. **(b)** Speakers and listeners probably perceive sermon length differently (just as, say, students and lecturers have different perceptions of the length of a class period).

8.31. Recall the rule of thumb from Chapter 5: Use the Normal approximation if $np \geq 10$ and $n(1 - p) \geq 10$. We use p_0 (the value specified in H_0) to make our decision.
(a) No: $np_0 = 6$. (b) Yes: $np_0 = 18$ and $n(1 - p_0) = 12$. (c) Yes: $np_0 = n(1 - p_0) = 50$.
(d) No: $np_0 = 2$.

8.33. With $\hat{p} = 0.69$, $\text{SE}_{\hat{p}} \doteq 0.02830$ and the 95% confidence interval is 0.6345 to 0.7455.

8.35. We estimate $\hat{p} = \frac{594}{2533} \doteq 0.2345$, $\text{SE}_{\hat{p}} \doteq 0.00842$, and the 95% confidence interval is 0.2180 to 0.2510.

8.37. We estimate $\hat{p} = \frac{110}{125} = 0.88$, $\text{SE}_{\hat{p}} \doteq 0.02907$, and the 95% confidence interval is 0.8230 to 0.9370.

8.39. **(a)** For testing H_0: $p = 0.5$ versus H_a: $p \neq 0.5$, we have $\hat{p} = \frac{5067}{10000} = 0.5067$ and $\sigma_{\hat{p}} = \sqrt{(0.5)(0.5)/10000} = 0.005$, so $z = \frac{0.0067}{0.005} = 1.34$, for which $P = 0.1802$. This is not significant at $\alpha = 0.05$ (or even $\alpha = 0.10$). **(b)** $\text{SE}_{\hat{p}} = \sqrt{\hat{p}(1 - \hat{p})/10000} \doteq$ 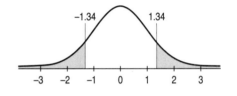 0.005, so the 95% confidence interval is $0.5067 \pm (1.96)(0.005)$, or 0.4969 to 0.5165.

8.41. As a quick estimate, we can observe that to cut the margin of error in half, we must quadruple the sample size, from 43 to 172. Using the sample-size formula, we find $n = \left(\frac{1.96}{2(0.075)}\right)^2 \doteq 170.7$—use $n = 171$. (The difference in the two answers is due to rounding.)

8.43. The required sample sizes are found by computing $\left(\frac{1.96}{0.1}\right)^2 p^*(1-p^*) = 384.16p^*(1-p^*)$: To be sure that we meet our target margin of error, we should take the largest sample indicated: $n = 97$ or larger.

p^*	n	Rounded up
0.1	34.57	35
0.2	61.47	62
0.3	80.67	81
0.4	92.20	93
0.5	96.04	97
0.6	92.20	93
0.7	80.67	81
0.8	61.47	62
0.9	34.57	35

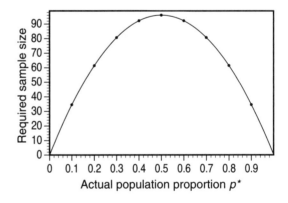

8.45. With $p_1 = 0.4$, $n_1 = 25$, $p_2 = 0.5$, and $n_2 = 30$, the mean and standard deviation of the sampling distribution of $D = \hat{p}_1 - \hat{p}_2$ are

$$\mu_D = p_1 - p_2 = -0.1 \text{ and } \sigma_D = \sqrt{\frac{p_1(1-p_1)}{n_1} + \frac{p_2(1-p_2)}{n_2}} \doteq 0.1339$$

8.47. (a) The means are $\mu_{\hat{p}_1} = p_1$ and $\mu_{\hat{p}_2} = p_2$. The standard deviations are

$$\sigma_{\hat{p}_1} = \sqrt{\frac{p_1(1-p_1)}{n_1}} \text{ and } \sigma_{\hat{p}_2} = \sqrt{\frac{p_2(1-p_2)}{n_2}}$$

(b) $\mu_D = \mu_{\hat{p}_1} - \mu_{\hat{p}_2} = p_1 - p_2$. **(c)** $\sigma_D^2 = \sigma_{\hat{p}_1}^2 + \sigma_{\hat{p}_2}^2 = \frac{p_1(1-p_1)}{n_1} + \frac{p_2(1-p_2)}{n_2}$.

8.49. Let us call the proportions favoring Commercial B q_w and q_m. Our estimates of these proportions are the complements of those found in Exercise 8.48; for example, $\hat{q}_w = \frac{56}{100} = 0.56 = 1 - \hat{p}_w$. Consequently, the standard error of the difference $\hat{q}_w - \hat{q}_m$ is the same as that for $\hat{p}_m - \hat{p}_w$: $\text{SE}_D = \sqrt{\frac{\hat{q}_w(1-\hat{q}_w)}{100} + \frac{\hat{q}_m(1-\hat{q}_m)}{140}} \doteq 0.06496$. The margin of error is therefore also the same, and the 95% confidence interval for $q_w - q_m$ is $(\hat{q}_w - \hat{q}_m) \pm (1.96)(0.06496) = -0.0030$ to 0.2516.

Note: *We followed the text's practice of subtracting the smaller proportion from the larger one.*

8.51. Because the sample proportions would tend to support the alternative hypothesis ($p_m > p_w$), the P-value is half as large ($P = 0.0288$), which would be enough to reject H_0 at the 5% level.

8.53. For H_0: $p_1 = p_2$, the pooled estimate of the proportion is $\hat{p} = \frac{198+295}{2822+1553} \doteq 0.1127$. The standard error is $\text{SE}_{D_p} \doteq 0.00999$, and the test statistic is $z = \frac{0.1198}{0.00999} \doteq 11.99$. The alternative hypothesis was not specified in this exercise; for either $p_1 \neq p_2$ or $p_1 < p_2$, the P-value associated with $z = 11.99$ would be tiny. (For the alternative $p_1 > p_2$, P would be nearly 1, and we would not reject H_0; however, it is hard to imagine why we would suspect that podcast use had decreased from 2006 to 2008.)

8.55. (a) The filled-in table is on the right. The values of X_1 and X_2 are estimated as $(0.54)(1063)$ and $(0.89)(1064)$. **(b)** The estimated

Population	Population proportion	Sample size	Count of successes	Sample proportion
1	p_1	1063	574	0.54
2	p_2	1064	947	0.89

difference is $\hat{p}_2 - \hat{p}_1 \doteq 0.35$.
(c) Large-sample methods should be appropriate, because we have large, independent samples from two populations. **(d)** With $\text{SE}_D \doteq 0.01805$, the 95% confidence interval is $0.35 \pm 0.03537 \doteq 0.3146$ to 0.3854. **(e)** The estimated difference is about 35%, and the interval is about 31.5% to 38.5%. **(f)** A possible concern is that adults were surveyed before Christmas, while teens were surveyed before and after Christmas. It might be that some of those teens may have received game consoles as gifts, but eventually grew tired of them.

8.57. (a) The filled-in table is on the right. The values of X_1 and X_2 are estimated as $(0.73)(1063)$ and $(0.76)(1064)$. **(b)** The estimated difference is $\hat{p}_2 - \hat{p}_1 \doteq 0.03$.

Population	Population proportion	Sample size	Count of successes	Sample proportion
1	p_1	1063	776	0.73
2	p_2	1064	809	0.76

(c) Large-sample methods should be appropriate, because we have large, independent samples from two populations. **(d)** With $SE_D \doteq 0.01889$, the 95% confidence interval is $0.03 \pm 0.03702 \doteq -0.0070$ to 0.0670. **(e)** The estimated difference is about 3%, and the interval is about -0.7% to 6.7%. **(f)** As in the solution to Exercise 8.55, a possible concern is that adults were surveyed before Christmas.

8.59. No; this procedure requires independent samples from different populations. We have one sample (of teens).

8.61. (a) H_0 should refer to p_1 and p_2 (population proportions) rather than $\hat{p}_1$ and $\hat{p}_2$ (sample proportions). **(b)** Knowing $\hat{p}_1 = \hat{p}_2$ does not tell us that the success counts are equal ($X_1 = X_2$) *unless* the sample sizes are equal ($n_1 = n_2$). **(c)** Confidence intervals only account for random sampling error.

8.63. Pet owners had the lower proportion of women, so we call them "population 2": $\hat{p}_2 = \frac{285}{595} \doteq 0.4790$. For non-pet owners, $\hat{p}_1 = \frac{1024}{1939} \doteq 0.5281$. $SE_D \doteq 0.02341$, so the 95% confidence interval is 0.0032 to 0.0950.

8.65. With equal sample sizes, the pooled estimate of the proportion is $\hat{p} = 0.255$, the average of $\hat{p}_1 = 0.29$ and $\hat{p}_2 = 0.22$. This can also be computed by taking $X_1 = (0.29)(1421) = 412.09$ and $X_2 = (0.22)(1421) = 312.62$, so $\hat{p} = (X_1 + X_2)/(1421 + 1421)$. The standard error for a significance test is $SE_{D_p} \doteq 0.01635$, and the test statistic is $z \doteq 4.28$ ($P < 0.0001$); we conclude that the proportions are different. The standard error for a confidence interval is $SE_D \doteq 0.01630$, and the 95% confidence interval is 0.0381 to 0.1019. The interval gives us an idea of how large the difference is: Music downloads dropped 4% to 10%.

8.67. (a) We find $\hat{p}_1 = \frac{73}{91} \doteq 0.8022$ and $\hat{p}_2 = \frac{75}{109} \doteq 0.6881$. For a confidence interval, $SE_D \doteq 0.06093$, so the 95% confidence interval for $p_1 - p_2$ is $(0.8022 - 0.6881) \pm (1.96)(0.06093) = -0.0053$ to 0.2335. **(b)** The question posed was, "Do high-tech companies tend to offer stock options more often than other companies?" Therefore, we test H_0: $p_1 = p_2$ versus H_a: $p_1 > p_2$. With $\hat{p}_1 \doteq 0.8022$, $\hat{p}_2 \doteq 0.6881$, and $\hat{p} = \frac{73+75}{91+109} = 0.74$, we find $SE_{D_p} \doteq 0.06229$, so $z = (\hat{p}_1 - \hat{p}_2)/SE_{D_p} \doteq 1.83$. This gives $P = 0.0336$. **(c)** We have fairly strong evidence that high-tech companies are more likely to offer stock options. However, the confidence interval tells us that the difference in proportions could be very small, or as large as 23%.

8.69. (a) $\hat{p}_f = \frac{48}{60} = 0.8$, so $\text{SE}_{\hat{p}} \doteq 0.05164$ for females. $\hat{p}_m = \frac{52}{132} = 0.3\overline{9}$, so $\text{SE}_{\hat{p}} \doteq 0.04253$ for males. (b) $\text{SE}_D = \sqrt{0.05164^2 + 0.04253^2} \doteq 0.06690$, so the interval is $(\hat{p}_f - \hat{p}_m) \pm 1.645\,\text{SE}_D$, or 0.2960 to 0.5161. There is (with high confidence) a considerably higher percent of juvenile references to females than to males.

8.71. We test H_0: $p_1 = p_2$ versus H_a: $p_1 \neq p_2$. With $\hat{p}_1 \doteq 0.5281$, $\hat{p}_2 \doteq 0.4790$, and $\hat{p} = \frac{1024+285}{1939+595} \doteq 0.5166$, we find $\text{SE}_{D_p} \doteq 0.02342$, so $z = (\hat{p}_1 - \hat{p}_2)/\text{SE}_{D_p} \doteq 2.10$. This gives $P = 0.0360$—significant evidence (at the 5% level) that a higher proportion of non-pet owners are women.

8.73. (a) While there is only a 5% chance of any interval being wrong, we have six (roughly independent) chances to make that mistake. (b) For 99.2% confidence, use $z^* = 2.65$. (Using software, $z^* \doteq 2.6521$, or 2.6383 using the exact value of $0.05/6 = 0.008\overline{3}$). (c) The margin of error for each interval is $z^*\text{SE}_{\hat{p}}$, so each interval is about 1.35 times wider than in the previous exercise. (If intervals are rounded to three decimal places, as on the right, the results are the same regardless of the value of z^* used.)

Genre	Interval
Racing	0.705 to 0.775
Puzzle	0.684 to 0.756
Sports	0.643 to 0.717
Action	0.632 to 0.708
Adventure	0.622 to 0.698
Rhythm	0.571 to 0.649

8.75. (a) The proportion is $\hat{p} = 0.164$, so $X = (0.164)(15,000) = 2460$ of the households in the sample are wireless only. (b) $\text{SE}_{\hat{p}} \doteq 0.00302$, so the 95% confidence interval is $0.164 \pm 0.00593 = 0.1581$ to 0.1699. (c) The estimate is 16.4%, and the interval is 15.8% to 17.0%. (d) The difference in the sample proportions is $D = 0.164 - 0.042 = 0.122$. (e) $\text{SE}_D \doteq 0.00344$, so the margin of error is $1.96\,\text{SE}_D \doteq 0.00674$. (The confidence interval is therefore 0.1153 to 0.1287.)

8.77. With $\hat{p}_1 = 0.43$, $\hat{p}_2 = 0.32$, and $n_1 = n_2 = 1430$, we have $\text{SE}_D \doteq 0.01799$, so the 95% confidence interval is $(0.43 - 0.32) \pm 0.03526 = 0.0747$ to 0.1453.

8.79. (a) and (b) The revised confidence intervals and z statistics are in the table below. (c) While the interval and z statistic change slightly, the conclusions are roughly the same. **Note:** *Even if the second sample size were as low as 100, the two proportions would be significantly different, albeit less so ($z = 2.15$, $P = 0.0313$).*

n_2	SE_D	m.e.	Interval	$\hat{p}$	SE_{D_p}	z	P
1000	0.01972	0.03866	0.0713 to 0.1487	0.3847	0.02006	5.48	< 0.0001
2000	0.01674	0.03281	0.0772 to 0.1428	0.3659	0.01668	6.59	< 0.0001

8.81. Student answers will vary. Shown on the right is the margin of error arising for sample sizes ranging from 500 to 2300; a graphical summary is not shown, but a good choice would be a plot of margin of error versus sample size.

n	m.e.
500	0.04035
800	0.03190
1100	0.02721
1430	0.02386
1700	0.02188
2000	0.02018
2300	0.01881

8.83. With $\hat{p}_m = 0.59$ and $\hat{p}_w = 0.56$, the standard error is $\text{SE}_D \doteq 0.03053$, the margin of error for 95% confidence is $1.96\text{SE}_D \doteq 0.05983$, and the confidence interval for $p_m - p_w$ is -0.0298 to 0.0898.

8.85. **(a)** People have different symptoms; for example, not all who wheeze consult a doctor. **(b)** In the table (below), we find for "sleep" that $\hat{p}_1 = \frac{45}{282} \doteq 0.1596$ and $\hat{p}_2 = \frac{12}{164} \doteq 0.0732$, so the difference is $\hat{p}_1 - \hat{p}_2 \doteq 0.0864$. Therefore, $\text{SE}_D \doteq 0.02982$ and the margin of error for 95% confidence is 0.05844. Other computations are performed in like manner. **(c)** It is reasonable to expect that the bypass proportions would be higher—that is, we expect more improvement where the pollution decreased—so we could use the alternative $p_1 > p_2$. **(d)** For "sleep," we find $\hat{p} = \frac{45+12}{282+164} \doteq 0.1278$ and $\text{SE}_{D_p} \doteq 0.03279$. Therefore, $z \doteq (0.1596 - 0.0732)/\text{SE}_{D_p} \doteq 2.64$. Other computations are similar. Only the "sleep" difference is significant. **(e)** 95% confidence intervals are shown below. **(f)** Part (b) showed improvement relative to control group, which is a better measure of the effect of the bypass, because it allows us to account for the improvement reported over time even when no change was made.

Complaint	Bypass minus congested				Bypass	
	$\hat{p}_1 - \hat{p}_2$	95% CI	z	P	$\hat{p}$	95% CI
Sleep	0.0864	0.0280 to 0.1448	2.64	0.0042	0.1596	0.1168 to 0.2023
Number	0.0307	−0.0361 to 0.0976	0.88	0.1897	0.1596	0.1168 to 0.2023
Speech	0.0182	−0.0152 to 0.0515	0.99	0.1600	0.0426	0.0190 to 0.0661
Activities	0.0137	−0.0395 to 0.0670	0.50	0.3100	0.0925	0.0586 to 0.1264
Doctor	−0.0112	−0.0796 to 0.0573	−0.32	0.6267	0.1174	0.0773 to 0.1576
Phlegm	−0.0220	−0.0711 to 0.0271	−0.92	0.8217	0.0474	0.0212 to 0.0736
Cough	−0.0323	−0.0853 to 0.0207	−1.25	0.8950	0.0575	0.0292 to 0.0857

8.87. With $\hat{p} = 0.56$, $\text{SE}_{\hat{p}} \doteq 0.01433$, so the margin of error for 95% confidence is $1.96\text{SE}_{\hat{p}} \doteq 0.02809$.

8.89. With $\hat{p}_1 = \frac{2}{3}$ and $\hat{p}_2 = 0.2$, we have $\hat{p} = \frac{134\hat{p}_1 + 237\hat{p}_2}{134 + 237} \doteq 0.3686$, $\text{SE}_{D_p} \doteq 0.05214$, and $z = 8.95$—very strong evidence of a difference. (If we assume that "two-thirds of the die-hard fans" and "20% of the less loyal fans" mean 89 and 47 fans, respectively, then $\hat{p} \doteq 0.3666$ and $z \doteq 8.94$; the conclusion is the same.) For a 95% confidence interval, $\text{SE}_D \doteq 0.04831$ and the interval is 0.3720 to 0.5613. (With $X_1 = 89$ and $X_2 = 47$, the interval is 0.3712 to 0.5606.)

8.91. The proportions, z-values, and P-values are:

Text	1	2	3	4	5	6	7	8	9	10
$\hat{p}$	.8718	.9000	.5372	.6738	.9348	.6875	.6429	.6471	.7097	.8759
z	4.64	6.69	0.82	5.31	5.90	5.20	3.02	2.10	6.60	9.05
P	≈ 0	≈ 0	.4133	≈ 0	≈ 0	≈ 0	.0025	.0357	≈ 0	≈ 0

We reject H_0: $p = 0.5$ for all texts except Text 3 and (perhaps) Text 8. If we are using a "multiple comparisons" procedure such as Bonferroni (see Chapter 6), we also might fail to reject H_0 for Text 7.

The last three texts do not seem to be any different from the first seven; the gender of the author does not seem to affect the proportion.

8.95. The difference becomes more significant (i.e., the P-value decreases) as the sample size increases. For small sample sizes, the difference between $\hat{p}_1 = 0.5$ and $\hat{p}_2 = 0.4$ is not significant, but with larger sample sizes, we expect that the sample proportions should be better estimates of their respective population proportions, so $\hat{p}_1 - \hat{p}_2 = 0.1$ suggests that $p_1 \neq p_2$.

n	z	P
40	0.90	0.3681
50	1.01	0.3125
80	1.27	0.2041
100	1.42	0.1556
400	2.84	0.0045
500	3.18	0.0015
1000	4.49	0.0000

8.97. (a) Using either trial and error, or the formula derived in part (b), we find that at least $n = 342$ is needed. **(b)** Generally, the margin of error is $m = z^* \sqrt{\dfrac{\hat{p}_1(1 - \hat{p}_1)}{n} + \dfrac{\hat{p}_2(1 - \hat{p}_2)}{n}}$; with $\hat{p}_1 = \hat{p}_2 = 0.5$, this is $m = z^* \sqrt{0.5/n}$. Solving for n, we find $n = (z^*/m)^2/2$.

8.99. (a) $p_0 = \frac{143,611}{181,535} \doteq 0.7911$. **(b)** $\hat{p} = \frac{339}{870} \doteq 0.3897$, $\sigma_{\hat{p}} \doteq 0.0138$, and $z = (\hat{p} - p_0)/\sigma_{\hat{p}} \doteq -29.1$, so $P \doteq 0$ (regardless of whether H_a is $p > p_0$ or $p \neq p_0$). This is very strong evidence against H_0; we conclude that Mexican Americans are underrepresented on juries. **(c)** $\hat{p}_1 = \frac{339}{870} \doteq 0.3897$, while $\hat{p}_2 = \frac{143,611 - 339}{181,535 - 870} \doteq 0.7930$. Then $\hat{p} \doteq 0.7911$ (the value of p_0 from part (a)), $\text{SE}_{D_p} \doteq 0.01382$, and $z \doteq -29.2$—and again, we have a tiny P-value and reject H_0.

8.101. In each case, the standard error is $\sqrt{\hat{p}(1 - \hat{p})/1280}$. One observation is that, while many feel that loans are a burden and wish they had borrowed less, a majority are satisfied with the benefits they receive from their education.

	$\hat{p}$	$\text{SE}_{\hat{p}}$	95% confidence interval
Burdened by debt	0.555	0.01389	0.5278 to 0.5822
Would borrow less	0.544	0.01392	0.5167 to 0.5713
More hardship	0.343	0.01327	0.3170 to 0.3690
Loans worth it	0.589	0.01375	0.5620 to 0.6160
Career opportunities	0.589	0.01375	0.5620 to 0.6160
Personal growth	0.715	0.01262	0.6903 to 0.7397

Chapter 9 Solutions

9.1. (a) The conditional distributions are given in the table below. For example, given that Explanatory $= 1$, the distribution of the response variable is $\frac{70}{200} = 35\%$ "Yes" and $\frac{130}{200} = 65\%$ "No." **(b)** The graphical display might take the form of a bar graph like the one shown below, but other presentations are possible. **(c)** One notable feature is that when Explanatory $= 1$, "No" is more common, but "Yes" and "No" are closer to being evenly split when Explanatory $= 2$.

Response	Explanatory variable	
variable	1	2
Yes	35%	45%
No	65%	55%

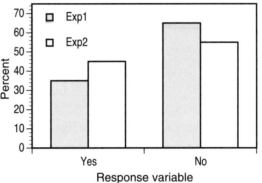

9.3. The relative risk is $\frac{0.00753}{0.00899} = 0.838$. Example 9.7 gave the 95% confidence interval $(1.02, 1.32)$, so with the ratio reversed, the interval would be approximately $(0.758, 0.980)$. For this relative risk, the statement made in Example 9.7 would be (changes underlined): "Since this interval does not include the value 1, corresponding to equal proportions in the two groups, we conclude that the <u>lower</u> CVD rate is statistically significant with $P < 0.05$. The <u>low</u> salt diet is associated with a <u>16% lower</u> rate of CVD than the <u>high</u> salt diet."

9.5. The table below summarizes the bounds for the P-values, and also gives the exact P-values (given by software). In each case, df $= (r - 1)(c - 1)$.

	X^2	Size of table	df	Crit. values (Table F)	Bounds for P	Actual P
(a)	5.32	2 by 2	1	$5.02 < X^2 < 5.41$	$0.02 < P < 0.025$	0.0211
(b)	2.7	2 by 2	1	$2.07 < X^2 < 2.71$	$0.10 < P < 0.15$	0.1003
(c)	25.2	4 by 5	12	$24.05 < X^2 < 26.22$	$0.01 < P < 0.02$	0.0139
(d)	25.2	5 by 4	12	$24.05 < X^2 < 26.22$	$0.01 < P < 0.02$	0.0139

9.7. The expected counts were rounded to the nearest hundredth.

9.9. The expected counts are in the table on the following page, rounded to four decimal places as in Example 9.15; for example, for California, we have

$$\frac{(257 - 269.524)^2}{269.524} \doteq 0.5820$$

The six values add up to 0.93 (rounded to two decimal places).

State	AZ	CA	HI	IN	NV	OH
Observed count	167	257	257	297	107	482
Proportion	0.105	0.172	0.164	0.188	0.070	0.301
Expected count	164.535	269.524	256.988	294.596	109.690	471.667
Chi-square contribution	0.0369	0.5820	0.0000	0.0196	0.0660	0.2264

9.11. (a) The two-way table is on the right; for example, for April 2001, $(0.05)(2250) = 112.5$ and $(0.95)(2250) = 2137.5$. **(b)** Under the null hypothesis that the proportions have not changed, the expected counts are

	Date of Survey			
Broadband?	April 2001	April 2004	March 2007	April 2008
Yes	112.5	540	1080	1237.5
No	2137.5	1710	1170	1012.5

$(0.33)(2250) = 742.5$ (across the top row) and $(0.67)(2250) = 1507.5$ (across the bottom row), because the average of the four broadband percents is $\frac{5\% + 24\% + 48\% + 55\%}{4} = 33\%$. (We take the unweighted average because we have assumed that the sample sizes were equal.) The test statistic is $X^2 \doteq 1601.8$ with df $= 3$, for which $P < 0.0001$. Not surprisingly, we reject H_0. **(c)** The average of the last two broadband percents is $\frac{48\% + 55\%}{2} = 51.5\%$, so if the proportions are equal, the expected counts are $(0.515)(2250) = 1158.75$ (top row) and $(0.485)(2250) = 1091.25$ (bottom row). The test statistic is $X^2 \doteq 22.07$ with df $= 1$, for which $P < 0.0001$.

Note: *This test is equivalent to testing H_0: $p_1 = p_2$ versus H_a: $p_1 \neq p_2$ using the methods of Chapter 8. We find pooled estimate $\hat{p} = 0.515$, $SE_{D_p} \doteq 0.01490$, and $z = (0.48 - 0.55)/SE_{D_p} \doteq -4.70$. (Note that $z^2 = X^2$.)*

9.13. Students may experiment with a variety of scenarios, but they should find that regardless of the what they try, the conclusion is the same.

9.15. (a) The 3×2 table is on the right. **(b)** The percents of disallowed small, medium, and large claims are (respectively) $\frac{6}{57} \doteq 10.5\%$, $\frac{5}{17} \doteq 29.4\%$, and $\frac{1}{5} = 20\%$. **(c)** In the 3×2 table, the expected count for large/not allowed is too small $(\frac{5 \cdot 12}{79} \doteq 0.76)$. **(d)** The null hypothesis is "There is no relationship between claim size and whether a claim is allowed."

	Allowed?		
Stratum	Yes	No	Total
Small	51	6	57
Medium	12	5	17
Large	4	1	5
Total	67	12	79

(e) As a 2×2 table (with the second row 16 "yes" and 6 "no"), we find $X^2 = 3.456$, df $= 1$, $P = 0.063$. The evidence is not quite strong enough to reject H_0.

9.17. The table on the right shows the given information translated into a 3×2 table. For example, in Year 1, about $(0.423)(2408) = 1018.584$ students received DFW grades, and the rest—$(0.577)(2408) = 1389.416$ students—passed.

Year	DFW	Pass
1	1018.584	1389.416
2	578.925	1746.075
3	423.074	1702.926

To test H_0: the DFW rate has not changed, we have $X^2 \doteq 307.8$, df $= 2$, $P < 0.0001$—very strong evidence of a change.

9.19. (a) The approximate counts are shown on the right; for example, among those students in trades, $(0.34)(942) = 320.28$ enrolled right after high school, and $(0.66)(942) = 621.72$ enrolled later. **(b)** In addition to a chi-square test in part (c), students might note other things, such as: Overall, 39.4% of these students enrolled right after high school. Health is the most popular field, with about 38% of these students.

Field of study	Time of entry		
	Right after high school	Later	Total
Trades	320.28	621.72	942
Design	274.48	309.52	584
Health	2034	3051	5085
Media/IT	975.88	2172.12	3148
Service	486	864	1350
Other	1172.60	1082.40	2255
Total	5263.24	8100.76	13,364

(c) We have strong enough evidence to conclude that there is an association between field of study and when students enter college; the test statistic is $X^2 = 275.9$ (with unrounded counts) or 276.1 (with rounded counts), with df $= 5$, for which P is very small. A graphical summary is not shown; a bar chart would be appropriate.

9.21. (a) The approximate counts are shown on the right; for example, among those students in trades, $(0.2)(942) = 188.4$ relied on parents, family, or spouse, and $(0.8)(942) = 753.6$ did not. **(b)** We have strong enough evidence to conclude that there is an association between field of study and getting money from parents, family, or spouse; the test statistic is $X^2 = 544.0$ (with un-

Field	Parents, family, spouse		
	Yes	No	Total
Trades	188.4	753.6	942
Design	221.63	377.37	599
Health	1360.84	3873.16	5234
Media/IT	518.08	2719.92	3238
Service	248.04	1129.96	1378
Other	943	1357	2300
Total	3479.99	10211.01	13,691

rounded counts) or 544.8 (with rounded counts), with df $= 5$, for which P is very small. **(c)** Overall, 25.4% of these students relied on family support; students in media/IT and service fields were slightly less likely, and those in the design and "other" fields were slightly more likely. A bar graph would be a good choice for a graphical summary.

9.23. (a) Of the high exercisers, $\frac{151}{151+148} \doteq 50.5\%$ get enough sleep, and the rest (49.5%) do not. **(b)** Of the low exercisers, $\frac{115}{115+242} \doteq 32.2\%$ get enough sleep, and the rest (67.8%) do not. **(c)** Those who exercise more than the median are more likely to get enough sleep. **(d)** To test H_0: exercise and sleep are not associated, we have $X^2 \doteq 22.58$ with df $= 1$, for which P is very small. We have very strong evidence of an association.

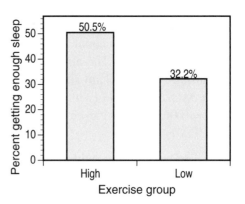

9.25. (a) The marginal totals are given in the table on the right. **(b)** The most appropriate description is the conditional distribution by gender (the explanatory variable): 91% of males, and 95% of females, agreed that trust and honesty are essential. **(c)** Females are slightly more likely to view trust and honesty as essential. **(d)** While the percents in the conditional distribution are similar, the large sample sizes make this highly significant: $X^2 \doteq 175.0$, df $= 1$, P is tiny. Once again, a P-value sketch is not shown.

Note: $X^2 = 175$ *coming from a $\chi^2(1)$ distribution is equivalent to $z = \sqrt{175} \doteq 13$ coming from the standard Normal distribution.*

Lied?	Male	Female	Total
Yes	11,724	14,169	25,893
No	1,163	746	1,909
Total	12,887	14,915	27,802

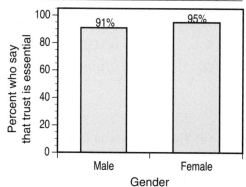

9.27. (a) The joint distribution is found by dividing each number in the table by 17,380 (the total of all the numbers). These proportions are given in italics on the right. For example, $\frac{3553}{17380} \doteq 0.2044$, meaning that about 20.4% of all college students are full-time and aged 15 to 19. **(b)** The marginal distribution of age is found by dividing the *row* totals by 17,380; they are in the right margin of the table (above, right) and the graph on the left below. For example, $\frac{3882}{17380} \doteq 0.2234$, meaning that about 22.3% of all college students are aged 15 to

	FT	PT	
15–19	3553	329	3882
	0.2044	*0.0189*	0.2234
20–24	5710	1215	6925
	0.3285	*0.0699*	0.3984
25–34	1825	1864	3689
	0.1050	*0.1072*	0.2123
35+	901	1983	2884
	0.0518	*0.1141*	0.1659
	11989	5391	17380
	0.6898	0.3102	

19. **(c)** The marginal distribution of status is found by dividing the *column* totals by 17,380; they are in the bottom margin of the table (above, right) and the graph on the right below. For example, $\frac{11989}{17380} \doteq 0.6898$, meaning that about 69% of all college students are full-time. **(d)** The conditional distributions are given in the table on the following page. For each status category, the conditional distribution of age is found by dividing the counts in that column by that column total. For example, $\frac{3553}{11989} \doteq 0.2964$, $\frac{5710}{11989} \doteq 0.4763$, etc., meaning that of all full-time college students, about 29.64% are aged 15 to 19, 47.63% are 20 to 24, and so on. Note that each set of four numbers should add to 1 (except for rounding error). Graphical presentations may vary; one possibility is shown on the following page. **(e)** We see that full-time students are dominated by younger ages, while part-time students are more likely to be older.

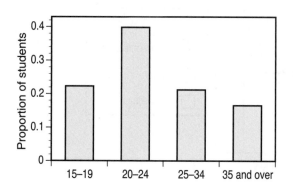

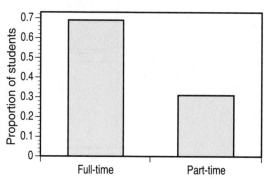

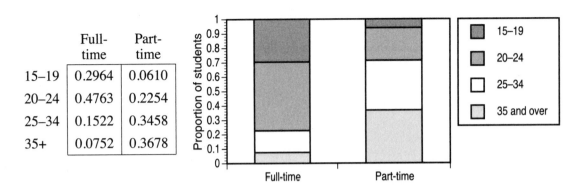

	Full-time	Part-time
15–19	0.2964	0.0610
20–24	0.4763	0.2254
25–34	0.1522	0.3458
35+	0.0752	0.3678

9.29. **(a)** The percent who have lasting waking symptoms is the total of the first column divided by the grand total: $\frac{69}{119} \doteq 57.98\%$. **(b)** The percent who have both waking and bedtime symptoms is the count in the upper left divided by the grand total: $\frac{36}{119} \doteq 30.25\%$. **(c)** To test H_0: There is no relationship between waking and bedtime symptoms versus H_a: There is a relationship, we find $X^2 \doteq 2.275$ (df $= 1$) and $P \doteq 0.132$. We do not have enough evidence to conclude that there is a relationship.

Minitab output

```
            WakeYes   WakeNo    Total
BedYes         36       33        69
             40.01    28.99

BedNo          33       17        50
             28.99    21.01

Total          69       50       119

ChiSq =   0.402 +  0.554 +
          0.554 +  0.765 = 2.275
df = 1, p = 0.132
```

9.31. Two examples are shown on the right. In general, choose a to be any number from 0 to 50, and then all the other entries can be determined.

30	20
70	80

10	40
90	60

> **Note:** *This is why we say that such a table has "one degree of freedom": We can make one (nearly) arbitrary choice for the first number, and then have no more decisions to make.*

9.33. **(a)** Different graphical presentations are possible; one is shown below. More women perform volunteer work; the notably higher percent of women who are "strictly voluntary" participants accounts for the difference. (The "court-ordered" and "other" percents are similar for men and women.) **(b)** Either by adding the three "participant" categories or by subtracting from 100% the non-participant percentage, we find that 40.3% of men and 51.3% of women are participants. The relative risk of being a volunteer is therefore $\frac{51.3\%}{40.3\%} \doteq 1.27$.

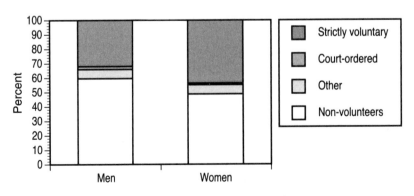

9.35. (a) The missing entries (shown shaded on the right) are found by subtracting the number who have tried low-fat diets from the given totals. **(b)** Viewing gender as explanatory, compute the conditional distributions of low-fat diet for each gender: $\frac{35}{181} \doteq 19.34\%$ of women and $\frac{8}{105} \doteq 7.62\%$ of men have tried low-fat diets. **(c)** The test statistic is $X^2 = 7.143$ (df = 1), for which $P = 0.008$. We have strong evidence of an association; specifically, women are more likely to try low-fat diets.

Low-fat diet?	Gender	
	Women	Men
Yes	35	8
No	146	97
Total	181	105

Minitab output

```
          Women      Men    Total
  Yes         35        8       43
           27.21    15.79

   No        146       97      243
          153.79    89.21

Total        181      105      286

ChiSq =   2.228 +  3.841 +
          0.394 +  0.680 = 7.143
df = 1, p = 0.008
```

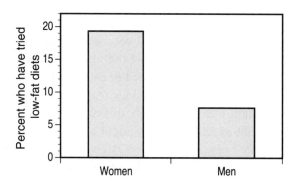

9.37. (a) As the conditional distribution of model dress for each age group has been given to us, it only remains to display this distribution graphically. One such presentation is shown below. **(b)** In order to perform the significance test, we must first recover the counts from the percents. For example, there were $(0.723)(1006) \doteq 727$ non-sexual ads in young adult magazines. The remainder of these counts can be seen in the Minitab output below, where we see $X^2 \doteq 2.59$, df = 1, and $P \doteq 0.108$—not enough evidence to conclude that age group affects model dress.

Minitab output

```
          Young   Mature    Total
     1      727      383     1110
          740.00   370.00

     2      279      120      399
          266.00   133.00

Total     1006      503     1509

ChiSq =   0.228 +  0.457 +
          0.635 +  1.271 = 2.591
df = 1, p = 0.108
```

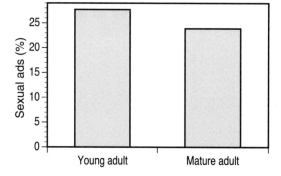

9.39. (a) First we must find the counts in each cell of the two-way table. For example, there were about $(0.172)(5619) \doteq 966$ Division I athletes who admitted to wagering. These counts are shown in the Minitab output on the right, where we see that $X^2 \doteq 76.7$, df $= 2$, and $P < 0.0001$. There is very strong evidence that the percent of athletes who admit to wa-

Minitab output

	Div1	Div2	Div3	Total
1	966	621	998	2585
	1146.87	603.54	834.59	
2	4653	2336	3091	10080
	4472.13	2353.46	3254.41	
Total	5619	2957	4089	12665

ChiSq = 28.525 + 0.505 + 31.996 +
         7.315 + 0.130 + 8.205 = 76.675
df = 2, p = 0.000

gering differs by division. **(b)** Even with much smaller numbers of students (say, 1000 from each division), P is still very small. Presumably the estimated numbers are reliable enough that we would not expect the true counts to be less than 1000, so we need not be concerned about the fact that we had to estimate the sample sizes. **(c)** If the reported proportions are wrong, then our conclusions may be suspect—especially if it is the case that athletes in some division were more likely to say they had not wagered when they had. **(d)** It is difficult to predict exactly how this might affect the results: Lack of independence could cause the estimated percents to be too large, or too small, if our sample included several athletes from teams which have (or do not have) a "gambling culture."

9.41. The Minitab output on the right shows both the two-way table (column and row headings have been changed to be more descriptive) and the results for the significance test: $X^2 \doteq 12.0$, df $= 1$, and $P = 0.001$, so we conclude that gender and flower choice are related. The count of 0 does not invalidate the test: Our smallest expected count is 6, while the text says that "for 2×2 tables, we require that all four expected cell counts be 5 or more."

Minitab output

	Female	Male	Total
bihai	20	0	20
	14.00	6.00	
no	29	21	50
	35.00	15.00	
Total	49	21	70

ChiSq = 2.571 + 6.000 +
         1.029 + 2.400 = 12.000
df = 1, p = 0.001

9.43. The graph on the right depicts the conditional distribution of pet ownership for each education level; for example, among those who did not finish high school, $\frac{421}{542} \doteq 77.68\%$ owned no pets, $\frac{93}{542} \doteq 17.16\%$ owned dogs, and $\frac{28}{542} \doteq 5.17\%$ (the rest) owned cats. (One could instead compute column percents—the conditional distribution of education for each pet-ownership group—but education level makes more sense as the explanatory variable here.) The (slightly altered) Minitab output shows that the relationship between education level and pet ownership is significant ($X^2 \doteq 23.15$, df = 4, $P < 0.0005$). Specifically, dog owners have less education, and cat owners more, than we would expect if there were no relationship between pet ownership and educational level.

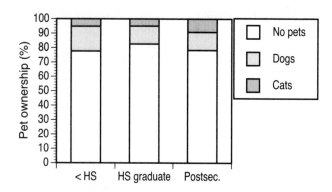

Minitab output

	None	Dogs	Cats	Total
<HS	421	93	28	542
	431.46	73.25	37.29	
HS	666	100	40	806
	641.61	108.93	55.46	
>HS	845	135	99	1079
	858.93	145.82	74.25	
Total	1932	328	167	2427

ChiSq = 0.253 + 5.326 + 2.316 +
        0.927 + 0.732 + 4.310 +
        0.226 + 0.803 + 8.254 = 23.147
df = 4, p = 0.000

9.45. The missing entries can be seen in the "Other" column of the Minitab output on the following page; they are found by subtracting the engineering, management, and liberal arts counts from each row total. The graph on the right shows the conditional distribution of transfer area for each initial major; for example, of those initially majoring in biology, $\frac{13}{398} \doteq 3.27\%$

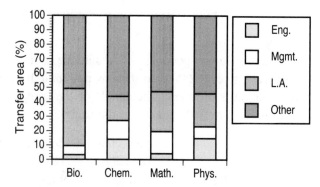

transferred to engineering, $\frac{25}{398} \doteq 6.28\%$ transferred to management, and so on. The relationship is significant ($X^2 \doteq 50.53$, df = 9, $P < 0.0005$). The largest contributions to X^2 come from chemistry or physics to engineering and biology to liberal arts (more transfers than expected) and biology to engineering and chemistry to liberal arts (fewer transfers than expected).

Minitab output

	Eng	Mgmt	LA	Other	Total
Bio	13	25	158	202	398
	25.30	34.56	130.20	207.95	
Chem	16	15	19	64	114
	7.25	9.90	37.29	59.56	
Math	3	11	20	38	72
	4.58	6.25	23.55	37.62	
Phys	9	5	14	33	61
	3.88	5.30	19.96	31.87	
Total	41	56	211	337	645

$$\text{ChiSq} = 5.979 + 2.642 + 5.937 + 0.170 +$$
$$10.574 + 2.630 + 8.973 + 0.331 +$$
$$0.543 + 3.608 + 0.536 + 0.004 +$$
$$6.767 + 0.017 + 1.777 + 0.040 = 50.527$$
$$\text{df} = 9, \ p = 0.000$$

9.47. The Minitab output on the right shows the 2×2 table and significance test details: $X^2 = 852.433$, $\text{df} = 1$, $P < 0.0005$. Using $z = -29.2$, computed in the solution to Exercise 8.81(c), this equals z^2 (up to rounding).

Minitab output

	Mex-Am	Other	Total
Juror	339	531	870
	688.25	181.75	
Not	143272	37393	180665
	142922.75	37742.25	
Total	143611	37924	181535

$$\text{ChiSq} = 177.226 + 671.122 +$$
$$0.853 + 3.232 = 852.433$$
$$\text{df} = 1, \ p = 0.000$$

9.49. The chi-square goodness of fit statistic is $X^2 \doteq 3.7807$ with $\text{df} = 3$, for which $P > 0.25$ (software gives 0.2861), so there is not enough evidence to conclude that this university's distribution is different. The details of the computation are given in the table below; note that there were 210 students in the sample.

	Expected frequency	Expected count	Observed count	$O - E$	$\dfrac{(O - E)^2}{E}$
Never	0.43	90.3	79	−11.3	1.4141
Sometimes	0.35	73.5	83	9.5	1.2279
Often	0.15	31.5	36	4.5	0.6429
Very often	0.07	14.7	12	−2.7	0.4959
			210		3.7807

9.55. (a) Each quadrant accounts for one-fourth of the area, so we expect it to contain one-fourth of the 100 trees. **(b)** *Some* random variation would not surprise us; we no more expect exactly 25 trees per quadrant than we would expect to see exactly 50 heads when flipping a fair coin 100 times. **(c)** The table on the right shows the individual computations, from which we obtain $X^2 = 10.8$, df $= 3$, and $P = 0.0129$. We conclude that the distribution is not random.

Observed	Expected	$(o - e)^2/e$
18	25	1.96
22	25	0.36
39	25	7.84
21	25	0.64
100		10.8

Chapter 10 Solutions

10.1. The given model was $\mu_y = 43.4 + 2.8x$, with standard deviation $\sigma = 4.3$. **(a)** The slope is 2.8. **(b)** When x increases by 1, μ_y increases by 2.8. (Or equivalently, if x increases by 2, μ_y increases by 5.6, etc.) **(c)** When $x = 7$, $\mu_y = 43.4 + 2.8(7) = 63$. **(d)** Approximately 95% of observed responses would fall in the interval $\mu_y \pm 2\sigma = 63 \pm 2(4.3) = 63 \pm 8.6 = 54.4$ to 71.6.

10.3. Example 10.5 gives the confidence interval -0.969 to -0.341 for the slope β_1. Recall that slope is the change in y (i.e., $\widehat{BMI}$) when x (i.e., PA) changes by $+1$. **(a)** If PA increases by 1, we expect $\widehat{BMI}$ to change by β_1, so the 95% confidence interval for the change is -0.969 to -0.341—that is, a decrease of 0.341 to 0.969 kg/m^2. **(b)** If PA decreases by 1, we expect $\widehat{BMI}$ to change by $-\beta_1$, so the 95% confidence interval for the change is an increase of 0.341 to 0.969 kg/m^2. **(c)** If PA increases by 0.5, we expect $\widehat{BMI}$ to change by $0.5\beta_1$, so the 95% confidence interval for the change is a decrease of 0.1705 to 0.4845 kg/m^2.

10.5. **(a)** The plot on the following page suggests a linear increase. **(b)** The regression equation is $\hat{y} = -4566.24 + 2.3x$. **(c)** The fitted values and residuals are given in the table below. Squaring the residuals and summing gives 0.952, so the standard error is:

$$s = \sqrt{\frac{0.952}{n-2}} = \sqrt{0.3173} \doteq 0.5633$$

(d) Given x (the year), spending comes from a $N(\mu_y, \sigma)$ distribution, where $\mu_y = \beta_0 + \beta_1 x$. The estimates of β_0, β_1, and σ are $b_0 = -4566.24$, $b_1 = 2.3$, and $s \doteq 0.5633$.
(e) We first note that $\bar{x} = 2005$ and $\sum(x_i - \bar{x})^2 = 10$, so $SE_{b_1} = s/\sqrt{10} \doteq 0.1781$. We have df $= n - 2 = 3$, so $t^* = 3.182$, and the 95% confidence interval for β_1 is $b_1 \pm t^* SE_{b_1} \doteq 2.3 \pm 0.5667 \doteq 1.733$ to 2.867. This gives the rate of increase of R&D spending: between 1.733 and 2.867 billion dollars per year.

Year	Spending ($billions)	Fitted values	Residuals
2003	40.1	40.66	-0.56
2004	43.3	42.96	0.34
2005	45.8	45.26	0.54
2006	47.7	47.56	0.14
2007	49.4	49.86	-0.46

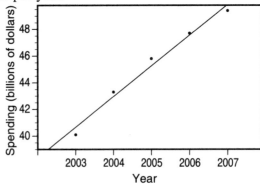

10.7. **(a)** The parameters are β_0, β_1, and σ; b_0, b_1, and s are the *estimates* of those parameters. **(b)** H_0 should refer to β_1 (the population slope) rather than b_1 (the estimated slope). **(c)** The confidence interval will be narrower than the prediction interval because the confidence interval accounts only for the uncertainty in our estimate of the mean response, while the prediction interval must also account for the random error of an individual response.

10.9. The test statistic is $t = b_1/SE_{b_1} = b_1/0.58$, with df $= n - 2$. The tests for parts (a) and (c) are not quite significant at the 5% level, while the test for part (b) is highly significant. This is consistent with the confidence intervals from the previous exercise.

	df	b_1	t	P (Table D)	P (software)
(a)	23	1.1	1.90	$0.05 < P < 0.10$	0.0705
(b)	23	2.1	3.62	$0.001 < P < 0.002$	0.0014
(c)	98	1.1	1.90	$0.05 < P < 0.10^*$	0.0608

*Note that for (c), if we use Table D, we take df $= 80$.

10.11. **(a)** In the Minitab output below, we have $SE_{b_1} \doteq 0.1604$. With df $= 30$, $t^* = 2.042$, so the 95% confidence interval for β_1 is $1.6924 \pm t^*SE_{b_1} \doteq 1.3649$ to 2.0199. This slope means that a \$1 difference in tuition in 2000 changes 2008 tuition by between \$1.36 and \$2.02. (It might be easier to understand expressed like this: If the costs of two schools differed by \$1000 in the year 2000, then in 2008, they would differ by between \$1365 and \$2020.) **(b)** Regression explains $r^2 \doteq 78.2\%$ of the variation in 2008 tuition. **(c)** When $x = 5100$, the estimated 2008 tuition is $\hat{y} = 1133 + 1.6924(5100) \doteq \$9764$. **(d)** When $x = 8700$, the estimated 2008 tuition is $\hat{y} = 1133 + 1.6924(8700) \doteq \$15{,}857$. (Software reports \$15,856; the difference is due to rounding. **(e)** The 2000 tuition at Stat U is similar to others in the data set, while Moneypit U was considerably more expensive in 2000, so that prediction requires extrapolation.

Minitab output: Regression of 2008 tuition on 2000 tuition

```
The regression equation is y2008 = 1133 + 1.69 y2000

Predictor      Coef        Stdev     t-ratio        p
Constant      1132.8        701.4       1.61     0.116
y2000         1.6924       0.1604      10.55     0.000

s = 1134        R-sq = 78.2%      R-sq(adj) = 77.5%
```

Estimates for Stat U and Moneypit U

```
    Fit   Stdev.Fit        95.0% C.I.            95.0% P.I.
   9764         245     (  9264,   10263)   (  7397,   12130)
  15856         749     ( 14329,   17384)   ( 13084,   18628) X
X denotes a row with very extreme X values
```

10.13. (a) The regression equation is $\hat{y} = -0.0127 + 0.0180x$, and $r^2 \doteq 80.0\%$. Not surprisingly, we find that BAC increases as beer consumption increases; the relationship is quite strong, with beer consumption explaining 80% of the variation in BAC. **(b)** To test H_0: $\beta_1 = 0$ versus H_a: $\beta_1 > 0$, we find $t = 7.48$ and $P < 0.0001$. There is very strong evidence that drinking more beers increases BAC. **(c)** The predicted mean BAC for $x = 5$ beers is 0.07712; the 90% prediction interval is 0.040 to 0.114. Steve might be safe, but cannot be sure that his BAC will be below 0.08.

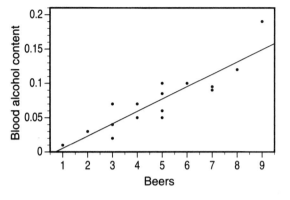

Note: *We use a prediction interval (rather than a confidence interval) because we want a range of values for an* individual *BAC after 5 beers, rather than the* mean *BAC.*

Minitab output: Regression of BAC on beer consumption

The regression equation is BAC = - 0.0127 + 0.0180 Beers

Predictor	Coef	Stdev	t-ratio	p
Constant	-0.01270	0.01264	-1.00	0.332
Beers	0.017964	0.002402	7.48	0.000

s = 0.02044   R-sq = 80.0%   R-sq(adj) = 78.6%

Fit	Stdev.Fit	90.0% C.I.	90.0% P.I.
0.07712	0.00513	(0.06808, 0.08616)	(0.03999, 0.11425)

10.15. (a) Both distributions are sharply right-skewed. Histograms and five-number summaries are on the following page. **(b)** Regression does not require that the variables x and y be Normal; it is the *errors* (the deviation from the line) that should be Normal. **(c)** The scatterplot is on the right; note that incentive pay is the explanatory variable. There is a weak positive linear relationship. **(d)** The regression equation is $\hat{y} = 6.247 + 0.1063x$. (Not surprisingly, regression only explains

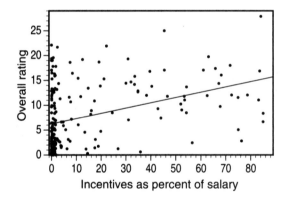

15.3% of the variation in rating.) **(e)** Residual analysis might include a histogram or stemplot, a plot of residuals versus incentive pay, and perhaps a Normal quantile plot; the first two of these items are shown on the following page. The residuals are slightly right-skewed. In addition, we note that for incentive pay less than about 30%, most residuals are greater than about -7, but extend up to $+15$.

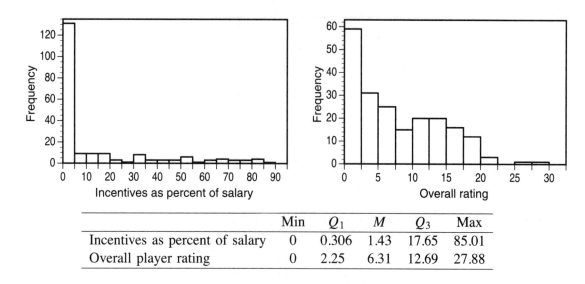

	Min	Q_1	M	Q_3	Max
Incentives as percent of salary	0	0.306	1.43	17.65	85.01
Overall player rating	0	2.25	6.31	12.69	27.88

Minitab output: Regression of rating on salary incentive percentage

```
The regression equation is Rating = 6.25 + 0.106 Percent

Predictor      Coef      Stdev    t-ratio       p
Constant      6.2469     0.4816     12.97    0.000
Percent       0.10634    0.01767     6.02    0.000

s = 5.854      R-sq = 15.3%      R-sq(adj) = 14.8%
```

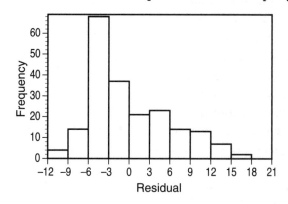
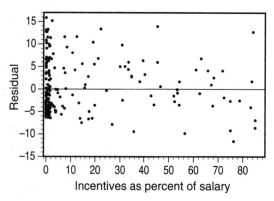

10.17. (a) 22 of these 30 homes sold for more than their assessed values. This was "an SRS of 30 properties," so it should be reasonably representative, so the larger population should be similar (or at least it was at the time of the sample). **(b)** The scatterplot (following page, left) shows a moderately strong linear association. **(c)** The regression line $\hat{y} = 21.50 + 0.9468x$ is included on the scatterplot. **(d)** The plot of residuals versus assessed value (following page, right) shows no obvious unusual features. The house with the highest assessed value (which also stands out in the original scatterplot) may be influential. **(e)** A stemplot (shown here) or histogram looks reasonably Normal, although there are two high residuals that stand apart from the rest. **(f)** There are no clear violations of the assumptions—at least, none severe enough to cause too much concern.

```
-3 | 3
-2 | 6543
-1 | 9422
-0 | 984430
 0 | 1134578
 1 | 1236
 2 | 26
 3 |
 4 | 17
```

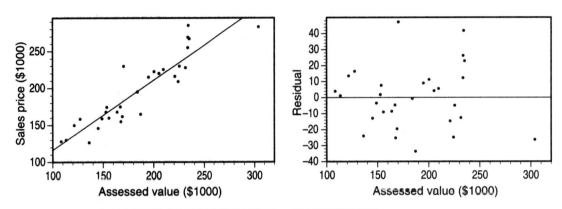

Minitab output: Regression of sales price on assessed value

```
The regression equation is SalesPrc = 21.5 + 0.947 Assessed

Predictor      Coef      Stdev    t-ratio        p
Constant      21.50      15.28       1.41    0.170
Assessed    0.94682    0.08064      11.74    0.000

s = 19.73      R-sq = 83.1%     R-sq(adj) = 82.5%
```

10.19. (a) The plot (below, left) is roughly linear and increasing. The number of tornadoes in 2004 (1819) is noticeably high, as is the 2008 count (1691) to a lesser extent. **(b)** The regression equation is $\hat{y} \doteq -28{,}438 + 14.82x$; both the slope and intercept are significantly different from 0. In the Minitab output on the following page, we see $SE_{b_1} \doteq 1.463$. With $t^* = 2.0049$ for df $= 54$, the confidence interval for β_1 is $b_1 \pm t^* SE_{b_1} = 14.82 \pm 2.93 \doteq 11.89$ to 17.76 tornadoes per year. **(c)** In the plot (below, right), we see that the scatter might

```
-3 | 520
-2 | 931
-1 | 99843310
-0 | 9887654443211110
 0 | 001223556778
 1 | 001224
 2 | 011789
 3 | 6
 4 |
 5 | 5
```

be greater in recent years, and the 2004 residual is particularly high. **(d)** Based on a stemplot (right), the 2004 residual is an outlier; the other residuals appear to be roughly Normal. **(e)** Without the 2004 count, the regression equation is $\hat{y} \doteq -26{,}584 + 13.88x$. The estimated slope decreases by almost one tornado per year.

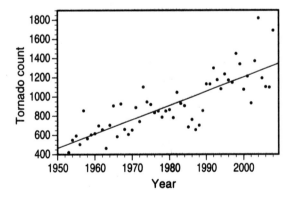

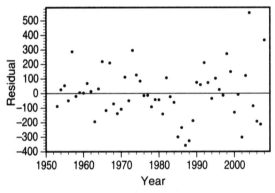

Minitab output: Regression of tornado count on year

The regression equation is Count = - 28438 + 14.8 Year

Predictor	Coef	Stdev	t-ratio	p
Constant	-28438	2897	-9.82	0.000
Year	14.822	1.463	10.13	0.000

s = 176.9   R-sq = 65.5%   R-sq(adj) = 64.9%

Regression with 2004 count removed

The regression equation is Count2 = - 26584 + 13.9 Year2

Predictor	Coef	Stdev	t-ratio	p
Constant	-26584	2680	-9.92	0.000
Year2	13.881	1.354	10.26	0.000

s = 160.5   R-sq = 66.5%   R-sq(adj) = 65.9%

10.21. (a) About $r^2 \doteq 8.41\%$ of the variability in AUDIT score is explained by (a linear regression on) gambling frequency. **(b)** With $r = 0.29$ and $n = 908$, the test statistic for H_0: $\rho = 0$ versus H_a: $\rho \neq 0$ is $t = \frac{r\sqrt{n-2}}{\sqrt{1-r^2}} \doteq 9.12$ (df = 906), for which P is very small.

(c) Nonresponse is a problem because the students who did not answer might have different characteristics from those who did. Because of this, we should be cautious about considering these results to be representative of all first-year students at this university, and even more cautious about extending these results to the broader population of all first-year students.

10.23. (a) The stemplots of percent forested and IBI are on the right. x (percent forested) is right-skewed; $\bar{x} = 39.3878\%$, $s_x = 32.2043\%$. y (IBI) is left-skewed; $\bar{y} = 65.9388$, $s_y = 18.2796$. **(b)** The scatterplot (following page, left) shows a weak positive association, with more scatter in y for small x. **(c)** $y_i = \beta_0 + \beta_1 x_i + \epsilon_i$, $i = 1, 2, ..., 49$; ϵ_i are independent $N(0, \sigma)$ variables. **(d)** The hypotheses are H_0: $\beta_1 = 0$ versus H_a: $\beta_1 \neq 0$. **(e)** See the Minitab output (following page). The regression equation is $\widehat{IBI} = 59.91 + 0.1531$ Forest, and the estimated standard deviation is $s \doteq 17.79$. For test-

Percent forested		IBI	
0	00000033789	2	99
1	0014778	3	233
2	125	3	9
3	123339	4	13
4	133799	4	67
5	229	5	34
6	38	5	556899
7	599	6	0124
8	069	6	7
9	055	7	11124
10	00	7	56889
		8	001222344
		8	556899
		9	1

ing the hypotheses in (d), $t = 1.92$ and $P = 0.061$. **(f)** The residual plot (following page, right) shows a slight curve—the residuals seem to be (very) slightly lower in the middle and higher on the ends. **(g)** As we can see from a stemplot and/or a Normal quantile plot (both on the following page), the residuals are left-skewed. **(h)** Student opinions may vary. The three apparent deviations from the model are (i) a possible change in standard deviation as x changes, (ii) possible curvature of residuals, and (iii) possible non-Normality of error terms.

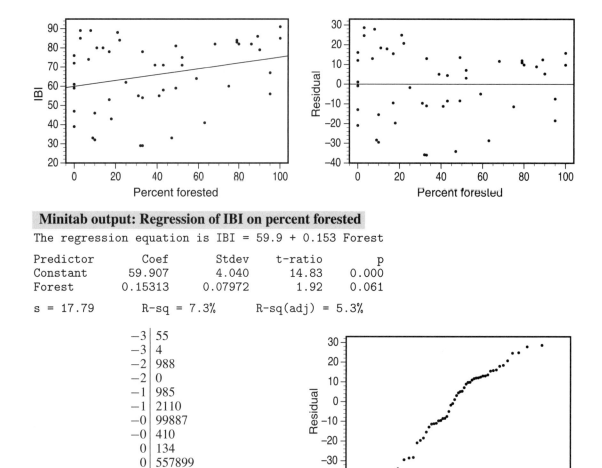

Minitab output: Regression of IBI on percent forested

The regression equation is IBI = 59.9 + 0.153 Forest

Predictor	Coef	Stdev	t-ratio	p
Constant	59.907	4.040	14.83	0.000
Forest	0.15313	0.07972	1.92	0.061

s = 17.79   R-sq = 7.3%   R-sq(adj) = 5.3%

```
-3 | 55
-3 | 4
-2 | 988
-2 | 0
-1 | 985
-1 | 2110
-0 | 99887
-0 | 410
 0 | 134
 0 | 557899
 1 | 01122333
 1 | 55678
 2 | 044
 2 | 78
```

10.25. The precise results of these changes depend on which observation is changed. (There are six observations which had 0% forest and two which had 100% forest.) Specifically, if we change IBI to 0 for one of the first six observations, the resulting P-value is between 0.019 (observation 6) and 0.041 (observation 3). Changing one of the last two observations changes the P-value to 0.592 (observation 48) or 0.645 (observation 49).

In general, the first change decreases P (that is, the relationship is more significant) because it accentuates the positive association. The second change weakens the association, so P increases (the relationship is less significant).

10.27. Using Area = 10 in the model $\widehat{IBI} = 52.92 + 0.4602$ Area from Exercise 10.22, $\widehat{IBI} \doteq 57.52$. Using Forest = 63 in the model $\widehat{IBI} = 59.91 + 0.1531$ Forest from Exercise 10.23, $\widehat{IBI} \doteq 69.55$. Both predictions have a lot of uncertainty; recall that r^2 was fairly small for both models. Also note that the prediction intervals (following page) are both about 70 units wide.

Minitab output: Predicting IBI for watershed area = 10					
Fit	Stdev.Fit	95.0% C.I.		95.0% P.I.	
57.52	3.41	(50.66,	64.39)	(23.55,	91.50)
Predicting IBI for percent forest = 63					
Fit	Stdev.Fit	95.0% C.I.		95.0% P.I.	
69.55	3.16	(63.19,	75.92)	(33.20,	105.91)

10.29. **(a)** Stemplots are shown below; both variables are right-skewed. For pure tones, $\bar{x} \doteq 106.20$ and $s \doteq 91.76$ spikes/second, and for monkey calls, $\bar{y} = 176.57$ and $s_y = 111.85$ spikes/second. **(b)** There is a moderate positive association; the third point (circled) has the largest residual; the first point (marked with a square) is an outlier for tone response. **(c)** With all 37 points, $\widehat{CALL} = 93.9 + 0.778\,TONE$ and $s = 87.30$; the test of $\beta_1 = 0$ gives $t = 4.91$, $P < 0.0001$. **(d)** Without the first point, $\hat{y} = 101 + 0.693x$, $s = 88.14$, $t = 3.18$. Without the third point, $\hat{y} = 98.4 + 0.679x$, $s = 80.69$, $t = 4.49$. With neither, $\hat{y} = 116 + 0.466x$, $s = 79.46$, $t = 2.21$. The line changes a bit, but always has a slope significantly different from 0.

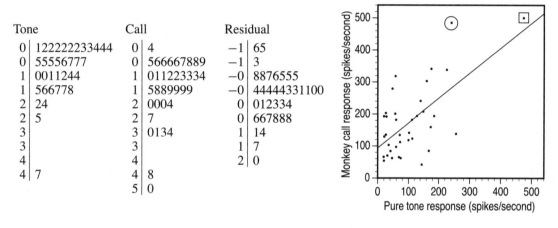

Tone	Call	Residual
0 | 122222233444	0 | 4	−1 | 65
0 | 55556777	0 | 566667889	−1 | 3
1 | 0011244	1 | 011223334	−0 | 8876555
1 | 566778	1 | 5889999	−0 | 44444331100
2 | 24	2 | 0004	0 | 012334
2 | 5	2 | 7	0 | 667888
3 |	3 | 0134	1 | 14
3 |	3 |	1 | 7
4 |	4 |	2 | 0
4 | 7	4 | 8	
	5 | 0	

10.31. **(a)** The stemplots (following page, left) are fairly symmetric. For x (MOE), $\bar{x} \doteq 1,799,180$ and $s_x \doteq 329,253$; for y (MOR), $\bar{y} \doteq 11,185$ and $s_y \doteq 1980$. **(b)** The plot (following page, right) shows a moderately strong, positive, linear relationship. Because we would like to predict MOR from MOE, we should put MOE on the x axis. **(c)** The model is $y_i = \beta_0 + \beta_1 x_i + \epsilon_i$, $i = 1, 2, ..., 32$; ϵ_i are independent $N(0, \sigma)$ variables. The regression equation is $\widehat{MOR} = 2653 + 0.004742\,MOE$, $s \doteq 1238$. The slope is significantly different from 0: $t = 7.02$ (df = 30), $P < 0.0001$. **(d)** Assumptions appear to be met: A stemplot of the residuals shows one slightly low (not quite an outlier), but acceptable, and the plot of residuals against MOE (not shown) does not suggest any particular pattern.

MOE	MOR	Residuals
11 6	6 3	−3 3
12	7	−2
13 55	8 3588	−2
14 1578	9 222	−1 6
15 5589	10 22356	−1 31110
16 14	11 223455799	−0 76555
17 2479	12 00777	−0 43221
18 447	13 469	0 00223
19 358	14 5	0 78
20 0348	15 3	1 1334
21 8		1 599
22 1		2 1
23 47		
24		
25 3		

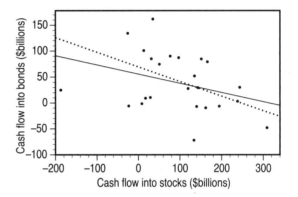

10.33. (a) The scatterplot shows a weak negative association; the regression equation is $\widehat{\text{Bonds}} = 55.58 - 0.1769\ \text{Stocks}$, with $s \doteq 54.55$. (This is the solid line in the plot.) **(b)** For testing $H_0\colon \beta_1 = 0$ versus $H_a\colon \beta_1 \neq 0$, we have $t = -1.66$ (df = 22) and $P = 0.111$. The slope is *not* significantly different from 0. **(c)** With the 2008 data removed, the (dashed) regression line is $\widehat{\text{Bonds}} = 69.46 - 0.2814\ \text{Stocks}$, with $s \doteq 53.12$. The slope is now significantly different from 0 ($t = -2.24$, $P = 0.036$). **(d)** We should explore whether something happened in 2008 that might explain why that point strayed from the line. (The economy would seem to be a likely cause.)

Minitab output: Regression of bond cash flow on stock cash flow

The regression equation is Bonds = 55.6 - 0.177 Stocks

Predictor	Coef	Stdev	t-ratio	p
Constant	55.58	14.98	3.71	0.001
Stocks	-0.1769	0.1066	-1.66	0.111

s = 54.55   R-sq = 11.1%   R-sq(adj) = 7.1%

Regression without 2008 data

The regression equation is Bonds = 69.5 - 0.281 Stocks

Predictor	Coef	Stdev	t-ratio	p
Constant	69.46	17.33	4.01	0.001
Stocks	-0.2814	0.1254	-2.24	0.036

s = 53.12   R-sq = 19.3%   R-sq(adj) = 15.5%

10.35. (a) Aside from the one high point (70 months of service, and wages 97.6801), there is a moderate positive association— fairly clear but with quite a bit of scatter. **(b)** The regression equation is $\widehat{\text{WAGES}} = 43.383 + 0.07325\text{ LOS}$, with $s \doteq 10.21$ (Minitab output below). The slope is significantly different from 0: $t = 2.85$ (df $= 57$), $P = 0.006$. **(c)** Wages rise an average of 0.07325 wage units per week of service. **(d)** We have $b_1 \doteq 0.07325$ and $SE_{b_1} \doteq 0.02571$. For a t distribution with df $= 57$, $t^* \doteq 2.0025$ for a 95% confidence interval, so the interval is 0.0218 to 0.1247.

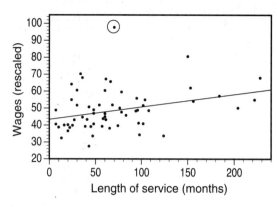

Minitab output: Regression of wages on length of service (outlier excluded)

The regression equation is wages = 43.4 + 0.0733 los

Predictor	Coef	Stdev	t-ratio	p
Constant	43.383	2.248	19.30	0.000
los	0.07325	0.02571	2.85	0.006

s = 10.21   R-sq = 12.5%   R-sq(adj) = 10.9%

Regression of wages on length of service (outlier included)

The regression equation is wages = 44.2 + 0.0731 los

Predictor	Coef	Stdev	t-ratio	p
Constant	44.213	2.628	16.82	0.000
los	0.07310	0.03015	2.42	0.018

s = 11.98   R-sq = 9.2%   R-sq(adj) = 7.6%

10.37. (a) The trend appears to be quite linear. **(b)** The regression equation is $\widehat{\text{Lean}} = -61.12 + 9.3187\,\text{Year}$ with $s \doteq 4.181$. The regression explains $r^2 = 98.8\%$ of the variation in lean. **(c)** The rate we seek is the slope. For df $= 11$ and 99% confidence, $t^* = 3.1058$, so the interval is $9.3187 \pm (3.1058)(0.3099) = 8.3562$ to 10.2812 tenths of a millimeter/year.

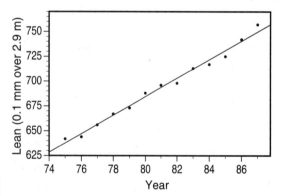

Minitab output: Regression of lean on year

The regression equation is Lean = -61.1 + 9.32 Year

Predictor	Coef	Stdev	t-ratio	p
Constant	-61.12	25.13	-2.43	0.033
Year	9.3187	0.3099	30.07	0.000

s = 4.181   R-sq = 98.8%   R-sq(adj) = 98.7%

10.39. (a) Use $x = 112$ (the number of years after 1900). **(b)** $\hat{y} = -61.12 + 9.3187(112) \doteq 983$, for a prediction of 2.9983 m. **(c)** A prediction interval is appropriate because we are interested in one future observation, not the mean of all future observations; in this situation, it does not make sense to talk of more than one future observation. In the output below, note that Minitab warns us of the risk of extrapolation.

Minitab output: Predicting lean in 2012 (year = 112)

```
    Fit  Stdev.Fit       95.0% C.I.            95.0% P.I.
  982.57       9.68  (  961.27, 1003.88)  (  959.36, 1005.78) XX
XX denotes a row with very extreme X values
```

10.41. To test H_0: $\rho = 0$ versus H_a: $\rho \neq 0$, we compute $t = \frac{r\sqrt{n-2}}{\sqrt{1-r^2}} \doteq -4.16$. Comparing this to a t distribution with df $= 116$, we find $P < 0.0001$, so we conclude the correlation is different from 0.

10.43. Recall that testing H_0: $\rho = 0$ versus H_a: $\rho \neq 0$ is the same as testing H_0: $\beta_1 = 0$ versus H_a: $\beta_1 \neq 0$. In the solution to Exercise 10.33, we had $t = -1.66$ (df $= 22$) and $P = 0.111$, so we cannot reject H_0.

10.45. For linear regression, DFM $= 1$. Because DFT $=$ DFM $+$ DFE and SST $=$ SSM $+$ SSE, we can find the missing degrees of freedom (DF) and sum of squares (SS) entries on

Source	DF	SS	MS	F
Regression	1	4560.6	4560.6	20.55
Residual	18	3995.4	221.97	
Total	19	8556.0		

the Residual row by subtraction: DFE $= 18$ and SSE $= 3995.4$. The entries in the mean square (MS) column are MSM $= \frac{SSM}{DFM} = 4560.6$ and MSE $= \frac{SSE}{DFE} \doteq 221.97$. Finally, $F = \frac{MSM}{MSE} \doteq 20.55$.

10.47. As $s_x = \sqrt{\frac{1}{19}\sum(x_i - \bar{x})^2} = 19.99\%$, we have $\sqrt{\sum(x_i - \bar{x})^2} = s_x\sqrt{19} \doteq 87.1344\%$, so:

$$SE_{b_1} = \frac{s}{\sqrt{\sum(x_i - \bar{x})^2}} \doteq \frac{14.8985}{87.1344} \doteq 0.1710$$

Alternatively, note that we have $F \doteq 20.55$ and $b_1 = 0.775$. Because $t^2 = F$, we know that $t = 4.5332$ (take the positive square root, because t and b_1 have the same sign). Then $SE_{b_1} = b_1/t = 0.1710$. (Note that with this approach, we do not need to know that $s_x = 19.99\%$.)

Finally, with df $= 18$, $t^* = 2.1009$ for 95% confidence, so the 95% confidence interval is $0.775 \pm 0.3592 = 0.4158$ to 1.1342.

10.49. Use the formula $t = \frac{r\sqrt{n-2}}{\sqrt{1-r^2}}$ with $r = 0.6$. For $n = 20$, $t = 3.18$ with df $= 18$, for which the two-sided P-value is $P = 0.0052$. For $n = 10$, $t = 2.12$ with df $= 8$, for which the two-sided P-value is $P = 0.0667$. With the larger sample size, r should be a better estimate of ρ, so we are less likely to get $r = 0.6$ unless ρ is really not 0.

10.51. (a) Not surprisingly, there is a positive association between scores. The 47th pair of scores (circled) is an outlier—the ACT score (21) is higher than one would expect for the SAT score (420). Since this SAT score is so low, this point may be influential. No other points fall outside the pattern. **(b)** The regression equation is $\hat{y} = 1.626 + 0.02137x$. The slope is significantly different from 0: $t = 10.78$ (df $= 58$) for which $P < 0.0005$. **(c)** $r = 0.8167$.

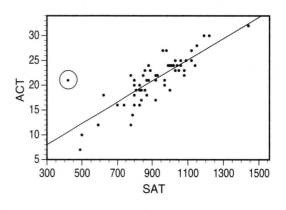

Minitab output: Regression of ACT score on SAT score

The regression equation is ACT = 1.63 + 0.0214 SAT

Predictor	Coef	Stdev	t-ratio	p
Constant	1.626	1.844	0.88	0.382
SAT	0.021374	0.001983	10.78	0.000

s = 2.744   R-sq = 66.7%   R-sq(adj) = 66.1%

10.53. (a) For SAT: $\bar{x} = 912.\overline{6}$ and $s_x = 180.1117$. For ACT: $\bar{y} = 21.1\overline{3}$ and $s_y = 4.7137$. Therefore, the slope is $a_1 \doteq 0.02617$ and the intercept is $a_0 \doteq -2.7522$. **(b)** The new line is dashed. **(c)** For example, the first prediction is $-2.7522 + (0.02617)(1000) \doteq 23.42$. Up to rounding error, the mean and standard deviation of the predicted scores are the same as those of the ACT scores: $\bar{y} = 21.1\overline{3}$ and $s_y = 4.7137$.

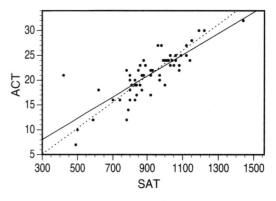

Note: *The usual least-squares line minimizes the total squared vertical distance from the points to the line. If instead we seek to minimize the total of $\sum_i |h_i v_i|$, where h_i is the horizontal distance and v_i is the vertical distance, we obtain the line $\hat{y} = a_0 + a_1 x$—except that we must choose the sign of a_1 to be the same as the sign of r. (It would hardly be the "best line" if we had a positive slope with a negative association.) If $r = 0$, either sign will do.*

10.55. (a) For squared length: $\widehat{\text{Weight}} = -117.99 + 0.4970$ SQLEN, $s \doteq 52.76$, $r^2 = 0.977$. **(b)** For squared width: $\widehat{\text{Weight}} = -98.99 + 18.732$ SQWID, $s \doteq 65.24$, $r^2 = 0.965$. Both scatterplots look more linear.

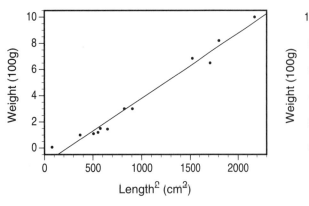

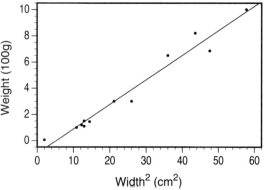

Minitab output: Regression of weight on squared length (Model 1)

```
The regression equation is weight = -118 + 0.497 sqlen

Predictor       Coef       Stdev    t-ratio        p
Constant      -117.99      27.88      -4.23    0.002
sqlen         0.49701     0.02400     20.71    0.000

s = 52.76     R-sq = 97.7%     R-sq(adj) = 97.5%
```

Regression of weight on squared width (Model 2)

```
The regression equation is weight = -99.0 + 18.7 sqwid

Predictor       Coef       Stdev    t-ratio        p
Constant       -98.99      33.67      -2.94    0.015
sqwid          18.732      1.126      16.64    0.000

s = 65.24     R-sq = 96.5%     R-sq(adj) = 96.2%
```

10.57. The table on the right shows the correlations and the corresponding test statistics. The first two results agree with the results of (respectively) Exercises 10.22 and 10.23.

	r	t	P
IBI/area	0.4459	3.42	0.0013
IBI/forest	0.2698	1.92	0.0608
area/forest	−0.2571	−1.82	0.0745

10.59. For each correlation, we compute $t = \frac{r\sqrt{n-2}}{\sqrt{1-r^2}}$. For the whole group, t ranges from 2.245 ($P = 0.0266$) to 3.208 ($P = 0.0017$). For Caucasians only, t ranges from 1.572 ($P = 0.1193$) to 2.397 ($P = 0.0185$). The three smallest correlations (0.16 and 0.19) are the only ones that are not significant.

Rule Breaking Measure	Popularity	Gene Expression
Sample 1 ($n = 123$)		
RB.composite	0.28**	0.26**
RB.questionnaire	0.22*	0.23*
RB.video	0.24**	0.20*
Sample 1 Caucasians only ($n = 96$)		
RB.composite	0.22*	0.23*
RB.questionnaire	0.16	0.24*
RB.video	0.19	0.16

10.61. (a) These intervals (in the table below) overlap quite a bit. **(b)** These quantities can be computed from the data, but it is somewhat simpler to recall that they can be found from the sample standard deviations $s_{x,\text{w}}$ and $s_{x,\text{m}}$:

$$s_{x,\text{w}}\sqrt{11} \doteq 6.8684\sqrt{11} \doteq 22.78 \quad \text{and} \quad s_{x,\text{m}}\sqrt{6} \doteq 6.6885\sqrt{6} \doteq 16.38$$

The women's SE_{b_1} is smaller in part because it is divided by a large number. **(c)** In order to reduce SE_{b_1} for men, we should choose our new sample to include men with a wider variety of lean body masses. (Note that just taking a larger sample will reduce SE_{b_1}; it is reduced even *more* if we choose subjects who will increase $s_{x,\text{m}}$.)

	b_1	SE_{b_1}	df	t^*	Interval
Women	24.026	4.174	10	2.2281	14.7257 to 33.3263
Men	16.75	10.20	5	2.5706	-9.4699 to 42.9699

Chapter 11 Solutions

11.1. (a) The response variable is math GPA. **(b)** The number of cases is $n = 106$. **(c)** There were $p = 4$ explanatory variables. **(d)** The explanatory variables were SAT Math, SAT Verbal, class rank, and mathematics placement score.

11.3. (a) The fact that the coefficients are all positive indicates that math GPA should increase when any explanatory variable increases (as we would expect). **(b)** With $n = 86$ cases and $p = 4$ variables, DFM $= p = 4$ and DFE $= n - p - 1 = 81$. **(c)** In the following table, each t statistic is the estimate divided by the standard error; the P-values are computed from a t distribution with df $= 81$. (The t statistic for the intercept was not required for this exercise, but is included for completeness.)

Variable	Estimate	SE	t	P
Intercept	-0.764	0.651	-1.1736	0.2440
SAT Math	0.00156	0.00074	2.1081	0.0381
SAT Verbal	0.00164	0.00076	2.1579	0.0339
HS rank	1.470	0.430	3.4186	0.0010
Bryant placement	0.889	0.402	2.2114	0.0298

All four coefficients are significantly different from 0 (although the intercept is not).

11.5. The correlations are found in Figure 11.4 and are summarized in the table on the right. Of the 15 possible scatterplots to be made from these six variables, three are shown below as examples. The pairs with the largest

	SATM	SATV	HSM	HSS	HSE
GPA	0.2517	0.1145	0.4365	0.3294	0.2890
SATM		0.4639	0.4535	0.2405	0.1083
SATV			0.2211	0.2617	0.2437
HSM				0.5757	0.4469
HSS					0.5794

correlations are generally easy to pick out. The whole-number scale for high school grades causes point clusters in those scatterplots and makes it difficult to determine the strength of the association. For example, in the plot of HSS versus HSE below, the circled point represents 9 of the 224 students. One might guess that these three scatterplots show relationships of roughly equal strength, but because of the overlapping points, the correlations are quite different; from left to right, they are 0.2517, 0.4365, and 0.5794.

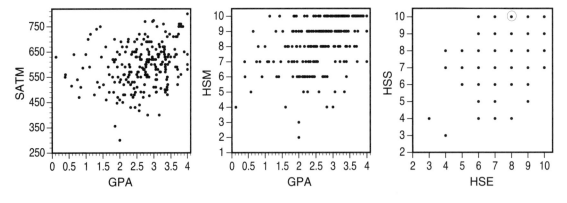

11.7. The table below gives two sets of answers: those found with critical values from Table D and those found with software. In each case, the estimated coefficient is $b_V = 6.4$ with standard error $SE_{b_1} = 3.1$, and the margin of error is $t^*SE_{b_1}$, with df $= n - 3$ for parts (a) and (b), and df $= n - 4$ for parts (c) and (d). (The Table D interval for part (d) uses df $= 100$.)

	n	df	t^*	Interval	t^*	Interval
			Table D		**Software**	
(a)	27	24	2.064	0.0016 to 12.7984	2.0639	0.0019 to 12.7981
(b)	53	50	2.009	0.1721 to 12.6279	2.0086	0.1735 to 12.6265
(c)	27	23	2.069	-0.0139 to 12.8139	2.0687	-0.0128 to 12.8128
(d)	124	120	1.984	0.2496 to 12.5504	1.9799	0.2622 to 12.5378

11.9. (a) H_0 should refer to β_2 (the population coefficient) rather than b_2 (the estimated coefficient). **(b)** This sentence should refer to the *squared* multiple correlation. **(c)** A small P implies that *at least one coefficient* is different from 0.

11.11. (a) $y_i = \beta_0 + \beta_1 x_{i1} + \beta_2 x_{i2} + \cdots + \beta_8 x_{i8} + \epsilon_i$, where $i = 1, 2, \ldots, 135$, and ϵ_i are independent $N(0, \sigma)$ random variables. **(b)** The sources of variation are model (DFM $= p = 8$), error (DFE $= n - p - 1 = 126$), and total (DFT $= n - 1 = 134$).

11.13. We have $p = 8$ explanatory variables and $n = 795$ observations. **(a)** The ANOVA F test has degrees of freedom DFM $= p = 8$ and DFE $= n - p - 1 = 786$. **(b)** This model explains only $R^2 \doteq 7.84\%$ of the variation in energy-drink consumption; it is not very predictive. **(c)** A positive (negative) coefficient means that large values of that variable correspond to higher (lower) energy-drink consumption. Therefore, males and Hispanics consume energy drinks more frequently, and consumption increases with risk-taking scores. **(d)** Within a group of students with identical (or similar) values of those other variables, energy-drink consumption increases with increasing jock identity and increasing risk taking.

11.15. We have $n = 202$, and $p = 1$ (for Model 1) or $p = 2$ (for Model 2). **(a)** For Model 1, DFE $= 200$. For Model 2, DFE $= 199$. **(b)** and **(c)** The test statistics $t = b_i/SE_{b_i}$ and P-values are in the

Model	Variable	t	P
1	Gene expression	$\frac{0.204}{0.066} \doteq 3.09$	0.0023
2	Gene expression	$\frac{0.161}{0.066} \doteq 2.44$	0.0153
	RB	$\frac{0.100}{0.030} \doteq 3.33$	0.0010

table on the right. **(d)** The relationship is still positive after adjusting for RB. When gene expression increases by 1, popularity increases by 0.204 in Model 1, and by 0.161 in Model 2 (with RB fixed).

11.17. (a) The regression equation is $\widehat{BMI} = 23.4 - 0.682x_1 + 0.102x_2$. (Minitab output on the following page.) **(b)** The quadratic regression explains $R^2 \doteq 17.7\%$ of the variation in BMI. **(c)** Analysis of residuals might include a stemplot, plots of residuals versus x_1 and x_2, and a Normal quantile plot. All of these appear on the following page; none suggest any obvious causes for concern. **(d)** From the Minitab output, $t = 1.83$ with df $= 97$, for which $P = 0.070$ — not significant.

Minitab output: Quadratic regression for predicting BMI from PA

```
The regression equation is BMI = 23.4 - 0.682 X1 + 0.102 X2

Predictor       Coef       Stdev     t-ratio        p
Constant      23.3956      0.4670      50.10     0.000
X1            -0.6818      0.1572      -4.34     0.000
X2            0.10195     0.05556       1.83     0.070

s = 3.612       R-sq = 17.7%      R-sq(adj) = 16.0%
```

```
-8 | 1
-7 | 3
-6 | 840
-5 | 9832
-4 | 553210
-3 | 8875421
-2 | 9943321
-1 | 7776311
-0 | 9776666642100
 0 | 0256667778889
 1 | 012334467899
 2 | 233447889
 3 | 67
 4 | 012277
 5 | 348
 6 | 68
 7 | 0007
```

11.19. **(a)** In the two scatterplots (below), we see a moderate positive linear relationship for small banks. For large banks, the relationship is very weak. **(b)** For small banks, $\widehat{\text{Wages}} = 35.9 + 0.1042\,\text{LOS}$, with $R^2 \doteq 46.6\%$ and $s \doteq 7.026$. **(c)** For large banks, $\widehat{\text{Wages}} = 49.5 + 0.0560\,\text{LOS}$, with $R^2 \doteq 3.5\%$ and $s \doteq 13.02$. **(d)** The large-bank regression is not significant (nor is it useful for prediction).

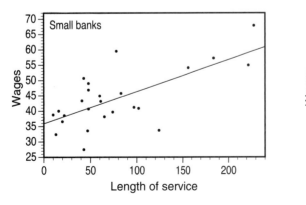

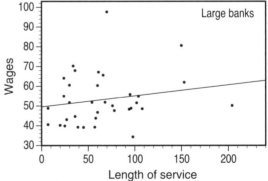

Minitab output: Regression of wages on length of service (small banks)

The regression equation is Wages-S = 35.9 + 0.104 LOS-S

Predictor	Coef	Stdev	t-ratio	p
Constant	35.914	2.284	15.73	0.000
LOS-S	0.10424	0.02328	4.48	0.000

s = 7.026   R-sq = 46.6%   R-sq(adj) = 44.3%

Regression of wages on length of service (large banks)

The regression equation is Wages-L = 49.5 + 0.0560 LOS-L

Predictor	Coef	Stdev	t-ratio	p
Constant	49.545	4.013	12.35	0.000
LOS-L	0.05595	0.05116	1.09	0.282

s = 13.02   R-sq = 3.5%   R-sq(adj) = 0.6%

11.21. (a) Budget and Opening are right-skewed; Theaters and Opinion are roughly symmetric (slightly left-skewed). Five-number summaries are best for skewed distributions, but all possible numerical summaries are given here.

Variable	$\bar{x}$	s	Min	Q_1	M	Q_3	Max
Budget	61.81	52.47	6.5	20.0	45.0	85.0	185.0
Opening	28.59	31.89	1.1	10.0	18.6	32.1	158.4
Theaters	2785	921	808.0	2123.0	2808.0	3510.0	4366.0
Opinion	6.440	1.064	3.6	5.9	6.6	7.0	8.9

A worthwhile observation is that for all four variables, the maximum observation comes from *The Dark Knight*. **(b)** Correlation coefficients are given with the scatterplots (below). All pairs of variables are positively correlated. The Budget/Theaters and Opening/Theaters relationships appear to be curved; the others are reasonably linear.

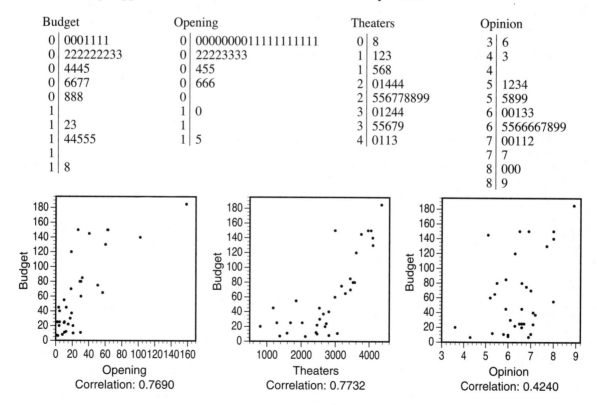

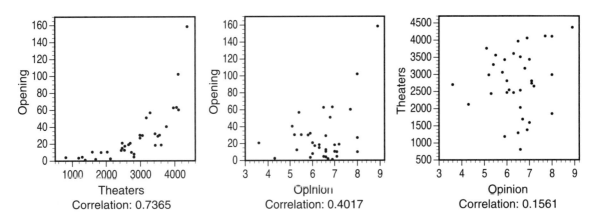

11.23. **(a)** The model is
$$\text{USRevenue}_i = \beta_0 + \beta_1 \, \text{Budget}_i + \beta_2 \, \text{Opening}_i + \beta_3 \, \text{Theaters}_i + \beta_4 \, \text{Opinion}_i + \epsilon_i$$
where $i = 1, 2, \ldots, 35$, and ϵ_i are independent $N(0, \sigma)$ random variables. **(b)** The regression equation is
$$\widehat{\text{USRevenue}} = -67.72 + 0.1351 \, \text{Budget} + 3.0165 \, \text{Opening}$$
$$- 0.00223 \, \text{Theaters} + 10.262 \, \text{Opinion}$$

(Minitab output below.) **(c)** On the following page is a stemplot of the residuals. The distribution is somewhat irregular, but a Normal quantile plot (not shown) does not suggest severe deviations from Normality. The residual analysis should also include a plot of residuals versus the explanatory variables; three of those plots are unremarkable (and not shown). The plot of residuals versus theaters suggests that the spread of the residuals increases with Theaters. *The Dark Knight*—noted as unusual in the previous two exercises—may be influential. **(d)** This regression explains $R^2 \doteq 98.1\%$ of the variation in revenue.

Minitab output: Regression of U.S. Revenue on budget, opening, theaters, and opinion

```
USRevenue = - 67.7 + 0.135 Budget + 3.02 Opening - 0.00223 Theaters
            + 10.3 Opinion
```

Predictor	Coef	Stdev	t-ratio	p
Constant	-67.72	24.14	-2.81	0.009
Budget	0.13511	0.09776	1.38	0.177
Opening	3.0165	0.1461	20.65	0.000
Theaters	-0.002229	0.005299	-0.42	0.677
Opinion	10.262	3.032	3.38	0.002

s = 15.69   R-sq = 98.1%   R-sq(adj) = 97.8%

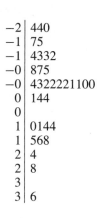

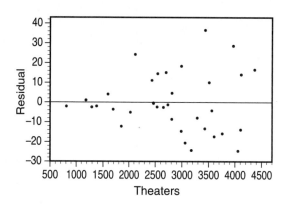

11.25. **(a)** Using the full model, the 95% prediction interval is \$86.86 to \$154.91 million. **(b)** With the reduced model, the interval is \$89.93 to \$155.00 million. **(c)** The intervals are very similar; as we saw in the previous exercise, there is little additional predictive information from the two variables we removed.

> **Note:** *According to* `http://www.imdb.com/title/tt0425061/business`, *the actual U.S. revenue for* Get Smart *was \$130.3 million.*

Minitab output: Predicting U.S. revenue for *Get Smart* (full model)			
Fit	Stdev.Fit	95.0% C.I.	95.0% P.I.
120.89	5.58	(109.48, 132.29)	(86.86, 154.91)
Predicting U.S. revenue for *Get Smart* (reduced model)			
Fit	Stdev.Fit	95.0% C.I.	95.0% P.I.
122.46	2.86	(116.64, 128.29)	(89.93, 155.00)

11.27. **(a)** The PEER distribution is left-skewed; the other two distributions are irregular (stemplots on the following page). Student choices of summary statistics may vary; both five-number summaries and means/standard deviations are given below. **(b)** Correlation coefficients are given below the scatterplots (following page). PEER and FtoS are negatively correlated, FtoS and CtoF are positively correlated, and the other correlation is very small.

Variable	$\bar{x}$	s	Min	Q_1	M	Q_3	Max
Peer review score	79.60	18.37	39	61	85	97	100
Faculty/student ratio	61.88	28.23	18	29	67	89	100
Citations/faculty ratio	63.84	25.23	17	40	66	86	100

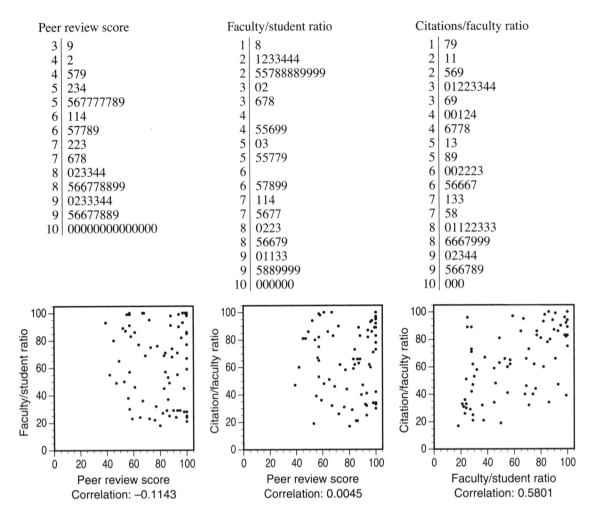

Peer review score
```
 3 | 9
 4 | 2
 4 | 579
 5 | 234
 5 | 567777789
 6 | 114
 6 | 57789
 7 | 223
 7 | 678
 8 | 023344
 8 | 566778899
 9 | 0233344
 9 | 56677889
10 | 00000000000000
```

Faculty/student ratio
```
 1 | 8
 2 | 1233444
 2 | 55788889999
 3 | 02
 3 | 678
 4 |
 4 | 55699
 5 | 03
 5 | 55779
 6 |
 6 | 57899
 7 | 114
 7 | 5677
 8 | 0223
 8 | 56679
 9 | 01133
 9 | 5889999
10 | 000000
```

Citations/faculty ratio
```
 1 | 79
 2 | 11
 2 | 569
 3 | 01223344
 3 | 69
 4 | 00124
 4 | 6778
 5 | 13
 5 | 89
 6 | 002223
 6 | 56667
 7 | 133
 7 | 58
 8 | 01122333
 8 | 6667999
 9 | 02344
 9 | 566789
10 | 000
```

Correlation: −0.1143   Correlation: 0.0045   Correlation: 0.5801

11.29. (a) The model is $OVERALL_i = \beta_0 + \beta_1 \, PEER_i + \beta_2 \, FtoS_i + \beta_3 \, CtoS_i + \epsilon_i$, where ϵ_i are independent $N(0, \sigma)$ random variables. **(b)** The regression equation is:

$$\widehat{OVERALL} = 18.85 + 0.5746 \, PEER + 0.0013 \, FtoS + 0.1369 \, CtoF$$

(c) For the confidence intervals, take $b_i \pm t^* SE_{b_i}$, with $t^* = 1.9939$ (for df $= 71$). These intervals have been added to the Minitab output below. The second interval contains 0, because that coefficient is not significantly different from 0. **(d)** The regression explains $R^2 \doteq 72.2\%$ of the variation in overall score. The estimate of σ is $s \doteq 7.043$.

Minitab output: Regression of overall score on all three variables

```
OVERALL = 18.8 + 0.575 PEER + 0.0013 FtoS + 0.137 CtoF
```

Predictor	Coef	Stdev	t-ratio	p	95% confidence interval
Constant	18.846	4.363	4.32	0.000	
PEER	0.57462	0.04504	12.76	0.000	0.4848 to 0.6644
FtoS	0.00130	0.03597	0.04	0.971	-0.0704 to 0.0730
CtoF	0.13690	0.03999	3.42	0.001	0.0572 to 0.2166

```
s = 7.043     R-sq = 72.2%     R-sq(adj) = 71.0%
```

11.31. **(a)** All distributions are skewed to varying degrees—GINI and CORRUPT to the right, the other three to the left. CORRUPT, DEMOCRACY, and LIFE have the most skewness. Student choices of summary statistics may vary; five-number summaries are a good choice because of the skewness, but some may also give means and standard deviations.

Variable	$\bar{x}$	s	Min	Q_1	M	Q_3	Max
LSI	6.2597	1.2773	2.5	5.5	6.1	7.35	8.2
GINI	37.9399	8.8397	24.70	32.65	35.95	42.750	60.10
CORRUPT	4.8861	2.4976	1.7	2.85	4.0	7.3	9.7
DEMOCRACY	4.2917	1.6799	0.5	3.0	5.0	5.5	6.0
LIFE	71.9450	9.0252	44.28	70.39	73.16	78.765	82.07

Notice especially how the skewness is apparent in the five-number summaries.

(b) Correlation coefficients are given below the scatterplots. GINI is negatively (and weakly) correlated to the other four variables, while all other correlations are positive and more substantial (0.533 or more).

LSI

```
2 | 5
3 | 3
3 | 78
4 |
4 | 55678
5 | 0023334
5 | 5555666777788889999
6 | 011233
6 | 557788
7 | 001113334
7 | 66677888899
8 | 000112
```

GINI

```
2 | 44
2 | 55566888999
3 | 0011223333344444444
3 | 55556666666788999
4 | 0001233344
4 | 56
5 | 0112234
5 | 778
6 | 0
```

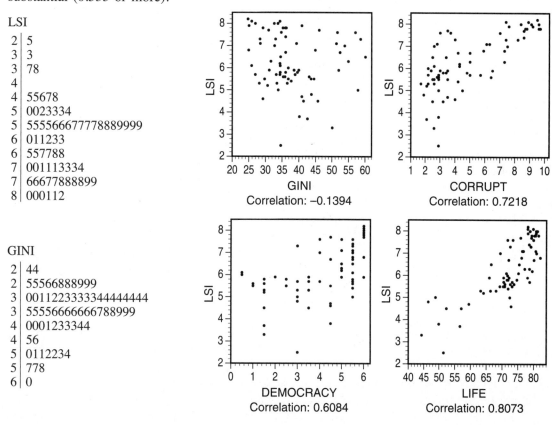

CORRUPT

```
1 | 79
2 | 112224
2 | 5556666688999999
3 | 002244
3 | 55557
4 | 002233
4 | 58
5 | 000
5 | 79
6 | 134
6 | 5
7 | 03344
7 | 56
8 | 24
8 | 66789
9 | 12
9 | 5667
```

DEMOCRACY

```
0 | 55
1 | 00
1 | 5555555
2 | 0
2 | 55
3 | 000000
3 | 5555
4 | 000
4 | 5555555
5 | 0000000
5 | 5555555555555555
6 | 000000000000000
```

LIFE

```
4 | 4
4 | 689
5 | 12
5 | 679
6 | 34
6 | 566899
7 | 0000001111112222223333 4
7 | 5556667888888889999999
8 | 0000000112
```

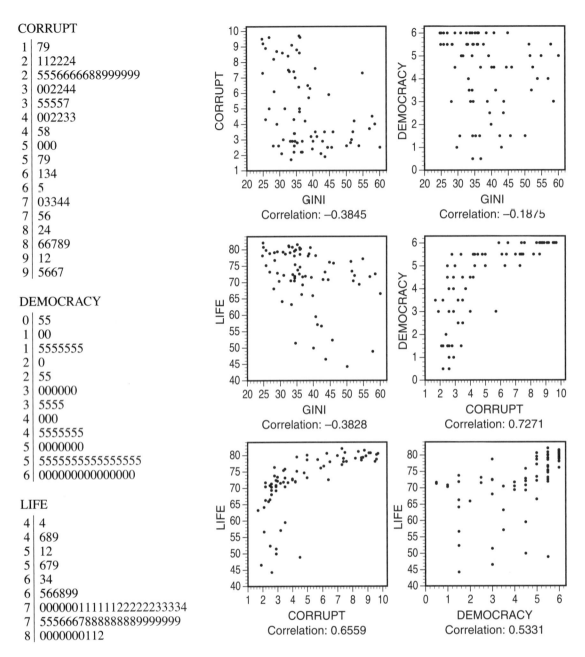

11.33. (a) The coefficients, standard errors, t statistics, and P-values are given in the Minitab output on the following page. **(b)** Student observations will vary. For example, the t statistic for the GINI coefficient grows from $t = -1.18$ ($P = 0.243$) to $t = 3.92$ ($P < 0.0005$). The DEMOCRACY t is 3.27 in the third model ($P < 0.0005$) but drops to 0.60 ($P = 0.552$) in the fourth model. **(c)** A good choice is to use GINI, LIFE, and CORRUPT. All three coefficients are significant, and $R^2 = 77.3\%$ is nearly the same as the fourth model from previous exercise. However, a scatterplot of the residuals versus CORRUPT (not shown) still looks quite a bit like the final scatterplot shown in the previous solution, suggesting a slightly curved relationship, which would violate the assumptions of our model.

Minitab output: Regression of LSI on GINI (Model 1)

LSI = 7.02 - 0.0201 GINI

Predictor	Coef	Stdev	t-ratio	p
Constant	7.0238	0.6660	10.55	0.000
GINI	-0.02014	0.01710	-1.18	0.243

s = 1.274   R-sq = 1.9%   R-sq(adj) = 0.5%

Regression of LSI on GINI and LIFE (Model 2)

LSI = - 3.83 + 0.0287 GINI + 0.125 LIFE

Predictor	Coef	Stdev	t-ratio	p
Constant	-3.8257	0.9746	-3.93	0.000
GINI	0.02873	0.01056	2.72	0.008
LIFE	0.12503	0.01034	12.09	0.000

s = 0.7266   R-sq = 68.6%   R-sq(adj) = 67.6%

Regression of LSI on GINI, LIFE, and DEMOCRACY (Model 3)

LSI = - 3.25 + 0.0280 GINI + 0.106 LIFE + 0.186 DEMOCRACY

Predictor	Coef	Stdev	t-ratio	p
Constant	-3.2524	0.9293	-3.50	0.001
GINI	0.028049	0.009891	2.84	0.006
LIFE	0.10634	0.01125	9.46	0.000
DEMOCRACY	0.18575	0.05682	3.27	0.002

s = 0.6804   R-sq = 72.8%   R-sq(adj) = 71.6%

Regression of LSI on all four variables (Model 4)

LSI = - 2.72 + 0.0368 GINI + 0.0905 LIFE + 0.0392 DEMOCRACY + 0.186 CORRUPT

Predictor	Coef	Stdev	t-ratio	p
Constant	-2.7201	0.8661	-3.14	0.003
GINI	0.036782	0.009393	3.92	0.000
LIFE	0.09048	0.01120	8.08	0.000
DEMOCRACY	0.03925	0.06566	0.60	0.552
CORRUPT	0.18554	0.05042	3.68	0.000

s = 0.6252   R-sq = 77.4%   R-sq(adj) = 76.0%

Regression of LSI on GINI, LIFE, and CORRUPT

LSI = - 2.74 + 0.0377 GINI + 0.0914 LIFE + 0.204 CORRUPT

Predictor	Coef	Stdev	t-ratio	p
Constant	-2.7442	0.8610	-3.19	0.002
GINI	0.037734	0.009213	4.10	0.000
LIFE	0.09141	0.01104	8.28	0.000
CORRUPT	0.20382	0.03991	5.11	0.000

s = 0.6222   R-sq = 77.3%   R-sq(adj) = 76.3%

11.35. (a) The scatterplot (following page, left) suggests greater variation in VO+ for large OC. The regression equation is

$$\widehat{\text{VO+}} = 334.0 + 19.505\,\text{OC}$$

with $s \doteq 443.3$ and $R^2 \doteq 0.435$; the test statistic for the slope is $t = 4.73$ ($P < 0.0005$), so we conclude the slope is not zero. The plot of residuals against OC suggests a slight downward curve on the right end, as well as increasing scatter as OC increases. The residuals are also somewhat right-skewed. A stemplot and Normal quantile plot of the residuals are not shown here but could be included as part of the analysis. **(b)** The regression equation is

$$\widehat{\text{VO+}} = 57.7 + 6.415\,\text{OC} + 53.87\,\text{TRAP}$$

with $s \doteq 376.3$ and $R^2 \doteq 0.607$. The coefficient of OC is not significantly different from 0 ($t = 1.25$, $P = 0.221$), but the coefficient of TRAP is ($t = 3.50$, $P = 0.002$). This is consistent with the correlations found in the solution to Exercise 11.34: TRAP is more highly correlated with VO+, and is also highly correlated with OC, so it is reasonable that, if TRAP is present in the model, little additional information is gained from OC.

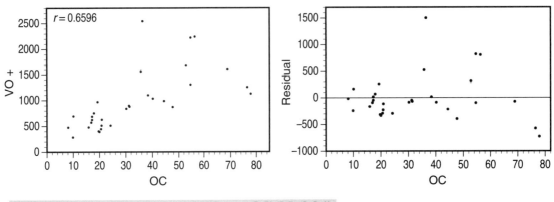

Minitab output: Regression of VO+ on OC (Model 1)

The regression equation is VOplus = 334 + 19.5 OC

Predictor	Coef	Stdev	t-ratio	p
Constant	334.0	159.2	2.10	0.045
OC	19.505	4.127	4.73	0.000

s = 443.3   R-sq = 43.5%   R-sq(adj) = 41.6%

Regression of VO+ on OC and TRAP (Model 2)

The regression equation is VOplus = 58 + 6.41 OC + 53.9 TRAP

Predictor	Coef	Stdev	t-ratio	p
Constant	57.7	156.5	0.37	0.715
OC	6.415	5.125	1.25	0.221
TRAP	53.87	15.39	3.50	0.002

s = 376.3   R-sq = 60.7%   R-sq(adj) = 57.9%

11.37. Stemplots (following page) show that all four variables are noticeably less skewed.

Variable	$\bar{x}$	s	Min	Q_1	M	Q_3	Max
LVO+	6.7418	0.5555	5.652	6.240	6.768	7.132	7.842
LVO–	6.6816	0.4832	5.537	6.284	6.806	6.935	7.712
LOC	3.3380	0.6085	2.092	2.885	3.408	3.865	4.355
LTRAP	2.4674	0.4978	1.194	2.175	2.332	2.944	3.360

Correlations and scatterplots (on the following page) show that all six pairs of variables are positively associated. The strongest association is between LVO+ and LVO– and the weakest is between LOC and LVO–. The regression equations for these transformed variables are given in the table that follows, along with significance test results. Residual analysis for these regressions is not shown.

The final conclusion is the same as for the untransformed data: When we use all three explanatory variables to predict LVO+, the coefficient of LTRAP is not significantly different from 0 and we then find that the model that uses LOC and LVO– to predict LVO+ is nearly as good (in terms of R^2), making it the best of the bunch.

LVO+	LVO−	LOC	LTRAP
5 | 6	5 | 5	2 | 0	1 | 1
5 | 99	5 |	2 | 23	1 |
6 | 011	5 | 8	2 |	1 |
6 | 223	6 | 001	2 | 7	1 | 7
6 | 4455	6 | 2223	2 | 8888999	1 | 89
6 | 67777	6 | 455	3 | 0001	2 | 001111
6 | 889	6 | 677	3 |	2 | 22233333
7 | 0011	6 | 8888888999	3 | 44455	2 |
7 | 33	7 | 01	3 | 667	2 | 6667
7 | 4	7 | 23	3 | 89	2 | 99999
7 | 77	7 | 4	4 | 000	3 | 1
7 | 8	7 | 7	4 | 233	3 | 223

	R^2	s

$$\widehat{\text{LVO+}} = 4.3841 + 0.7063\,\text{LOC} \qquad\qquad 0.599 \quad 0.3580$$

$$\begin{aligned}
&\text{SE} = 0.1074 \\
&t = 6.58 \\
&P < 0.0005
\end{aligned}$$

$$\widehat{\text{LVO+}} = 4.2590 + 0.4304\,\text{LOC} + 0.4240\,\text{LTRAP} \qquad 0.652 \quad 0.3394$$

$$\begin{aligned}
&\text{SE} = 0.1680 \qquad \text{SE} = 0.2054 \\
&t = 2.56 \qquad\quad\ t = 2.06 \\
&P = 0.016 \qquad\ P = 0.048
\end{aligned}$$

$$\widehat{\text{LVO+}} = 0.8716 + 0.3922\,\text{LOC} + 0.0275\,\text{LTRAP} + 0.6725\,\text{LVOminus} \quad 0.842 \quad 0.2326$$

$$\begin{aligned}
&\text{SE} = 0.1154 \qquad \text{SE} = 0.1570 \qquad \text{SE} = 0.1178 \\
&t = 3.40 \qquad\quad\ t = 0.18 \qquad\quad\ t = 5.71 \\
&P = 0.002 \qquad\ P = 0.842 \qquad\ P < 0.0005
\end{aligned}$$

$$\widehat{\text{LVO+}} = 0.8321 + 0.4061\,\text{LOC} \qquad\qquad\quad + 0.6816\,\text{LVOminus} \quad 0.842 \quad 0.2286$$

$$\begin{aligned}
&\text{SE} = 0.0824 \qquad\qquad\qquad\qquad \text{SE} = 0.1038 \\
&t = 4.93 \qquad\qquad\qquad\qquad\quad\ t = 6.57 \\
&P < 0.0005 \qquad\qquad\qquad\qquad P < 0.0005
\end{aligned}$$

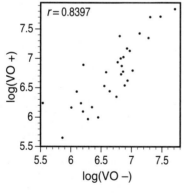

$r = 0.8397$

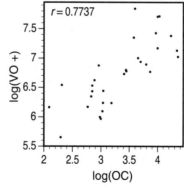

$r = 0.7737$

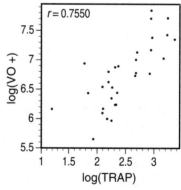

$r = 0.7550$

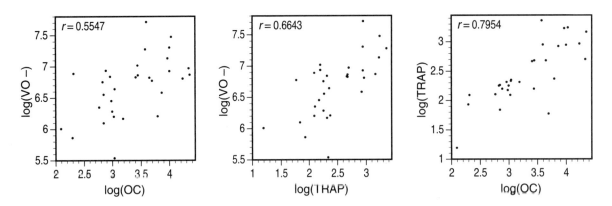

11.39. Refer to the solution to Exercise 11.37 for the scatterplots. As in the previous exercise, the more logical single-variable model would be to use LTRAP to predict LVO–, but many students might miss that detail. Both single-explanatory variable models are given in the following table. Residual analysis plots are not included. This time, we might conclude that the best model is to predict LVO– from LVO+ alone; neither biomarker variable makes an indispensable contribution to the prediction.

				R^2	s
$\widehat{\text{LVO}-} = 5.2110$	$+ 0.4406\,\text{LOC}$ $\text{SE} = 0.1227$ $t = 3.59$ $P = 0.001$			0.308	0.4089
$\widehat{\text{LVO}-} = 5.0905$		$+ 0.6449\,\text{LTRAP}$ $\text{SE} = 0.1347$ $t = 4.79$ $P < 0.0005$		0.441	0.3674
$\widehat{\text{LVO}-} = 5.0370$	$+ 0.0569\,\text{LOC}$ $\text{SE} = 0.1848$ $t = 0.31$ $P = 0.761$	$+ 0.5896\,\text{LTRAP}$ $\text{SE} = 0.2259$ $t = 2.61$ $P = 0.014$		0.443	0.3732
$\widehat{\text{LVO}-} = 1.5729$	$- 0.2932\,\text{LOC}$ $\text{SE} = 0.1407$ $t = -2.08$ $P = 0.047$	$+ 0.2447\,\text{LTRAP}$ $\text{SE} = 0.1662$ $t = 1.47$ $P = 0.152$	$+ 0.8134\,\text{LVOplus}$ $\text{SE} = 0.1425$ $t = 5.71$ $P < 0.0005$	0.748	0.2558
$\widehat{\text{LVO}-} = 1.3109$	$- 0.1878\,\text{LOC}$ $\text{SE} = 0.1237$ $t = -1.52$ $P = 0.140$		$+ 0.8896\,\text{LVOplus}$ $\text{SE} = 0.1355$ $t = 6.57$ $P < 0.0005$	0.728	0.2611
$\widehat{\text{LVO}-} = 1.7570$			$+ 0.7304\,\text{LVOplus}$ $\text{SE} = 0.0877$ $t = 8.33$ $P < 0.0005$	0.705	0.2669

11.41. (a) The model is:

$$PCB_i = \beta_0 + \beta_1\,PCB_{52} + \beta_2\,PCB_{118}$$
$$+ \beta_3\,PCB_{138} + \beta_4\,PCB_{180} + \epsilon_i$$

where $i = 1, 2, \ldots, 69$; ϵ_i are independent $N(0, \sigma)$ random variables. **(b)** The regression equation is:

$$\widehat{PCB} = 0.937 + 11.8727\,PCB_{52} + 3.7611\,PCB_{118}$$
$$+ \; 3.8842\,PCB_{138} + 4.1823\,PCB_{180}$$

```
-2 | 2
-1 |
-1 | 31
-0 | 8776655
-0 | 4443333222211111111000000
 0 | 000000000000011111222223333444
 0 | 677778
 1 | 12
 1 |
 2 | 2
```

with $s = 6.382$ and $R^2 = 0.989$. All coefficients are significantly different from 0, although the constant 0.937 is not ($t = 0.76$, $P = 0.449$). That makes some sense—if none of these four congeners are present, it might be somewhat reasonable to predict that the total amount of PCB is 0. **(c)** The residuals appear to be roughly Normal, but with two outliers. There are no clear patterns when plotted against the explanatory variables (these plots are not shown).

Minitab output: Regression of PCB on PCB52, PCB118, PCB138, and PCB180

```
PCB = 0.94 + 11.9 PCB52 + 3.76 PCB118 + 3.88 PCB138 + 4.18 PCB180
```

Predictor	Coef	Stdev	t-ratio	p
Constant	0.937	1.229	0.76	0.449
PCB52	11.8727	0.7290	16.29	0.000
PCB118	3.7611	0.6424	5.85	0.000
PCB138	3.8842	0.4978	7.80	0.000
PCB180	4.1823	0.4318	9.69	0.000

```
s = 6.382      R-sq = 98.9%     R-sq(adj) = 98.8%
```

11.43. (a) The regression equation is:

$$\widehat{PCB} = -1.018 + 12.644\,PCB_{52} + 0.3131\,PCB_{118} + 8.2546\,PCB_{138}$$

with $s = 9.945$ and $R^2 = 0.973$. Residual analysis (not shown) suggests a few areas of concern: The distribution of residuals has heavier tails than a Normal distribution, and the scatter (that is, prediction error) is greater for larger values of the predicted PCB. **(b)** The estimated coefficient of PCB118 is $b_2 \doteq 0.3131$; its P-value is 0.708. (Details in Minitab output below.) **(c)** In Exercise 11.41, $b_2 \doteq 3.7611$ and $P < 0.0005$. **(d)** This illustrates how complicated multiple regression can be: When we add PCB180 to the model, it complements PCB118, making it useful for prediction.

Minitab output: Regression of PCB on PCB52, PCB118, and PCB138

```
PCB = -1.02 + 12.6 PCB52 + 0.313 PCB118 + 8.25 PCB138
```

Predictor	Coef	Stdev	t-ratio	p
Constant	-1.018	1.890	-0.54	0.592
PCB52	12.644	1.129	11.20	0.000
PCB118	0.3131	0.8333	0.38	0.708
PCB138	8.2546	0.3279	25.18	0.000

```
s = 9.945      R-sq = 97.3%     R-sq(adj) = 97.2%
```

11.45. The model is:

$$TEQ_i = \beta_0 + \beta_1 \, PCB_{52} + \beta_2 \, PCB_{118}$$
$$+ \beta_3 \, PCB_{138} + \beta_4 \, PCB_{180} + \epsilon_i$$

where $i = 1, 2, \ldots, 69$; ϵ_i are independent $N(0, \sigma)$ random variables. The regression equation is:

$$\widehat{TEQ} = 1.0600 - 0.0973 \, PCB_{52} + 0.3062 \, PCB_{118}$$
$$+ 0.1058 \, PCB_{138} - 0.0039 \, PCB_{180}$$

-1	66
-1	4200
-0	987666666666555555
-0	44444333221111100
0	0000222224
0	566667788
1	23334
1	9
2	3
2	57

with $s = 0.9576$ and $R^2 = 0.677$. Only the constant and the PCB118 coefficient are significantly different from 0; see Minitab output below. Residuals (stemplot on the right) are slightly right-skewed and show no clear patterns when plotted with the explanatory variables (not shown).

Minitab output: Regression of TEQ on the four PCB congeners

```
TEQ = 1.06 - 0.097 PCB52 + 0.306 PCB118 + 0.106 PCB138 - 0.0039 PCB180

Predictor        Coef        Stdev    t-ratio        p
Constant       1.0600       0.1845       5.75    0.000
PCB52         -0.0973       0.1094      -0.89    0.377
PCB118        0.30618      0.09639       3.18    0.002
PCB138        0.10579      0.07470       1.42    0.162
PCB180       -0.00391      0.06478      -0.06    0.952

s = 0.9576     R-sq = 67.7%     R-sq(adj) = 65.7%
```

11.47. (a) The correlations (all positive) are listed in the table below. The largest correlation is 0.956 (LPCB and LPCB138); the smallest (0.227, for LPCB28 and LPCB180) is not quite significantly different from 0 ($t = 1.91$, $P = 0.0607$) but, with 28 correlations, such a P-value could easily arise by chance, so we would not necessarily conclude that $\rho = 0$. Rather than showing all 28 scatterplots—which are all fairly linear and confirm the positive associations suggested by the correlations—we have included only two of the interesting ones: LPCB against LPCB28 and LPCB against LPCB126. The former is notable because of one outlier (specimen 39) in LPCB28; the latter stands out because of the "stack" of values in the LPCB126 data set that arose from the adjustment of the zero terms. (The outlier in LPCB28 and the stack in LPCB126 can be seen in other plots involving those variables; the two plots shown are the most appropriate for using the PCB congeners to predict LPCB, as the next exercise asks.) **(b)** All correlations are higher with the transformed data. In part, this is because these scatterplots do not exhibit the "greater scatter in the upper right" that was seen in many of the scatterplots of the original data.

	LPCB28	LPCB52	LPCB118	LPCB126	LPCB138	LPCB153	LPCB180
LPCB52	0.795						
LPCB118	0.533	0.671					
LPCB126	0.272	0.331	0.739				
LPCB138	0.387	0.540	0.890	0.792			
LPCB153	0.326	0.519	0.780	0.647	0.922		
LPCB180	0.227	0.301	0.654	0.695	0.896	0.867	
LPCB	0.570	0.701	0.906	0.729	0.956	0.905	0.829

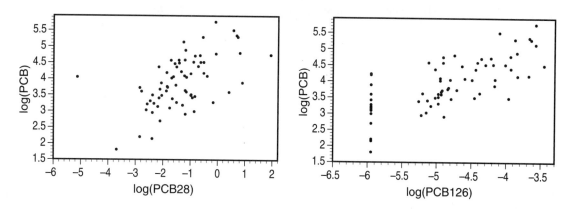

11.49. Using Minitab's BREG (best regression) command for guidance, we see that there is little improvement in R^2 beyond models with four explanatory variables. The best models with two, three, and four variables are given in the Minitab output below.

Minitab output: Best subsets regression

```
                                        L L L L L
                                    L L P P P P P
                                    P P C C C C C
                                    C C B B B B B
                                    B B 1 1 1 1 1
                     Adj.           2 5 1 2 3 5 8
Vars  R-sq  R-sq   C-p       s      8 2 8 6 8 3 0
   1  72.9  72.5  10.8  0.31266              X
   2  76.8  76.1   2.0  0.29166     X        X
   3  77.6  76.6   1.6  0.28859     X    X X
   4  78.0  76.7   2.5  0.28816     X    X X       X
   5  78.1  76.4   4.2  0.28981     X X X X        X
   6  78.2  76.1   6.1  0.29188     X X X X        X X
   7  78.2  75.7   8.0  0.29400     X X X X X X X
```

Best regression using two explanatory variables

The regression equation is
LTEQ = 3.96 + 0.107 LPCB28 + 0.622 LPCB126

Predictor	Coef	Stdev	t-ratio	p
Constant	3.9637	0.2275	17.42	0.000
LPCB28	0.10749	0.03242	3.32	0.001
LPCB126	0.62231	0.04801	12.96	0.000

s = 0.2917   R-sq = 76.8%   R-sq(adj) = 76.1%

Best regression using three explanatory variables

The regression equation is
LTEQ = 3.44 + 0.0777 LPCB28 + 0.114 LPCB118 + 0.543 LPCB126

Predictor	Coef	Stdev	t-ratio	p
Constant	3.4445	0.4029	8.55	0.000
LPCB28	0.07773	0.03736	2.08	0.041
LPCB118	0.11371	0.07319	1.55	0.125
LPCB126	0.54345	0.06952	7.82	0.000

s = 0.2886   R-sq = 77.6%   R-sq(adj) = 76.6%

Best regression using four explanatory variables

```
The regression equation is
LTEQ = 3.56 + 0.0720 LPCB28 + 0.170 LPCB118 + 0.554 LPCB126 - 0.0693 LPCB153

Predictor      Coef      Stdev    t-ratio       p
Constant     3.5568     0.4152       8.57   0.000
LPCB28      0.07199    0.03767       1.91   0.060
LPCB118     0.16973    0.08928       1.90   0.062
LPCB126     0.55374    0.07005       7.90   0.000
LPCB153    -0.06929    0.06344      -1.09   0.279

s = 0.2882      R-sq = 78.0%     R-sq(adj) = 76.7%
```

11.51. In the table, two *IQR*s are given; those in parentheses are based on quartiles reported by Minitab, which computes quartiles in a slightly different way from this text's method.

	$\bar{x}$	M	s	IQR
Taste	24.53	20.95	16.26	23.9 (or 24.58)
Acetic	5.498	5.425	0.571	0.656 (or 0.713)
H2S	5.942	5.329	2.127	3.689 (or 3.766)
Lactic	1.442	1.450	0.3035	0.430 (or 0.4625)

None of the variables show striking deviations from Normality in the quantile plots (not shown). Taste and H2S are slightly right-skewed, and Acetic has two peaks. There are no outliers.

```
Taste            Acetic            H2S             Lactic
0│ 00            4│ 455            2│ 9             8│ 6
0│ 556           4│ 67            3│ 1268899        9│ 9
1│ 1234          4│ 8             4│ 17799         10│ 689
1│ 55688         5│ 1             5│ 024           11│ 56
2│ 011           5│ 2222333       6│ 11679         12│ 5599
2│ 556           5│ 444           7│ 4699          13│ 013
3│ 24            5│ 677           8│ 7             14│ 469
3│ 789           5│ 888           9│ 025           15│ 2378
4│ 0             6│ 0011         10│ 1             16│ 38
4│ 7             6│ 3                              17│ 248
5│ 4             6│ 44                             18│ 1
5│ 67                                              19│ 09
                                                   20│ 1
```

11.53. The regression equation is $\widehat{\text{Taste}} = -61.50 + 15.648$ Acetic with $s = 13.82$ and $R^2 = 0.302$. The slope is significantly different from 0 ($t = 3.48$, $P = 0.002$).

Based on a stemplot (right) and quantile plot (not shown), the residuals seem to have a Normal distribution. Scatterplots (following page) reveal positive associations between residuals and both H2S and Lactic. The plot of residuals against Acetic suggests greater scatter in the residuals for large Acetic values.

```
-2│ 9
-2│ 11
-1│ 65
-1│ 31
-0│ 7655
-0│ 21
 0│ 0122224
 0│ 5668
 1│
 1│ 5679
 2│ 0
 2│ 6
```

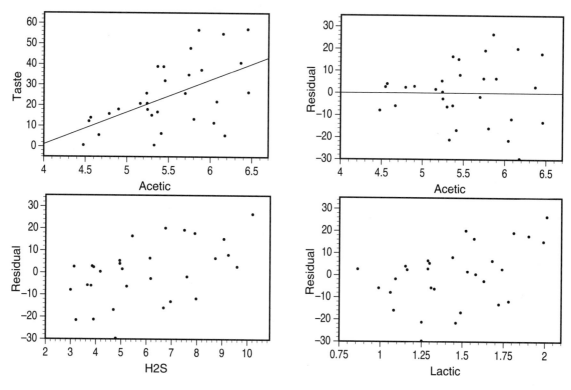

11.55. The regression equation is $\widehat{\text{Taste}} = -29.86 + 37.720$ Lactic with $s = 11.75$ and $R^2 = 0.496$. The slope is significantly different from 0 ($t = 5.25$, $P < 0.0005$).

Based on a stemplot (right) and quantile plot (not shown), the residuals appear to be roughly Normal. Scatterplots (below) reveal no striking patterns for residuals vs. Acetic and H2S.

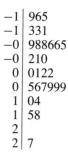

```
-1 | 965
-1 | 331
-0 | 988665
-0 | 210
 0 | 0122
 0 | 567999
 1 | 04
 1 | 58
 2 |
 2 | 7
```

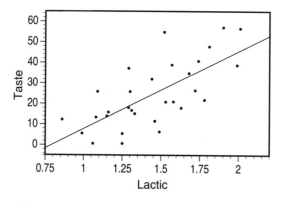

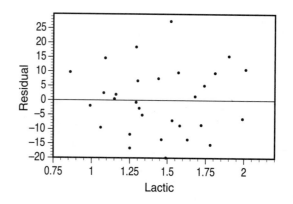

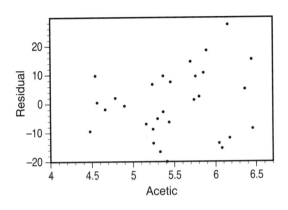

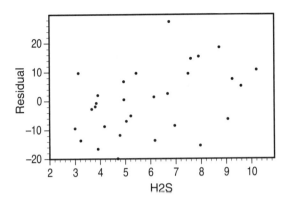

11.57. The regression equation is $\widehat{\text{Taste}} = -26.94 + 3.801$ Acetic $+ 5.146$ H2S with $s = 10.89$ and $R^2 = 0.582$. The t-value for the coefficient of Acetic is 0.84 ($P = 0.406$), indicating that it does not add significantly to the model when H2S is used because Acetic and H2S are correlated (in fact, $r = 0.618$ for these two variables). This model does a better job than any of the three simple linear regression models, but it is not much better than the model with H2S alone (which explained 57.1% of the variation in Taste)—as we might expect from the t-test result.

Minitab output: Regression of taste on acetic and h2s

```
taste = -26.9 + 3.80 acetic + 5.15 h2s

Predictor        Coef       Stdev     t-ratio          p
Constant       -26.94       21.19       -1.27      0.215
acetic          3.801       4.505        0.84      0.406
h2s             5.146       1.209        4.26      0.000

s = 10.89       R-sq = 58.2%      R-sq(adj) = 55.1%
```

11.59. The regression equation is $\widehat{\text{Taste}} = -28.88 + 0.328$ Acetic $+ 3.912$ H2S $+ 19.671$ Lactic with $s = 10.13$. The model explains 65.2% of the variation in Taste (the same as for the model with only H2S and Lactic). Residuals of this regression appear to be Normally distributed and show no patterns in scatterplots with the explanatory variables. (These plots are not shown.)

The coefficient of Acetic is not significantly different from 0 ($P = 0.942$); there is no gain in adding Acetic to the model with H2S and Lactic. It appears that the best model is the H2S/Lactic model of Exercise 11.58.

Minitab output: Regression of taste on acetic, h2s, and lactic

```
taste = -28.9 + 0.33 acetic + 3.91 h2s + 19.7 lactic

Predictor        Coef       Stdev     t-ratio          p
Constant       -28.88       19.74       -1.46      0.155
acetic          0.328       4.460        0.07      0.942
h2s             3.912       1.248        3.13      0.004
lactic         19.671       8.629        2.28      0.031

s = 10.13       R-sq = 65.2%      R-sq(adj) = 61.2%
```

Chapter 12 Solutions

12.1. (a) H_0 says the *population* means are all equal. **(b)** *Experiments* are best for establishing causation. **(c)** ANOVA is used to compare *means* (and assumes that the variances are equal). **(d)** Multiple comparisons procedures are used when we wish to determine which means are significantly different, but have no specific relations in mind before looking at the data. (Contrasts are used when we have prior expectations about the differences.)

12.3. We were given sample sizes $n_1 = 23$, $n_2 = 20$, and $n_3 = 28$ and standard deviations $s_1 = 5$, $s_2 = 5$, and $s_3 = 6$. **(a)** Yes: The guidelines for pooling standard deviations say that the ratio of largest to smallest should be less than 2; we have $\frac{6}{5} \doteq 1.2 < 2$. **(b)** Squaring the three standard deviations gives $s_1^2 = 25$, $s_2^2 = 25$, and $s_3^2 = 36$.
(c) $s_p^2 = \dfrac{22s_1^2 + 19s_2^2 + 27s_3^2}{22 + 19 + 27} \doteq 29.3676$. **(d)** $s_p = \sqrt{s_p^2} \doteq 5.4192$.

12.5. (a) This sentence describes *between*-group variation. Within-group variation is the variation that occurs by chance among members of the same group. **(b)** The *sums of* squares (not the mean squares) in an ANOVA table will add. **(c)** The common population standard deviation σ (not its estimate s_p) is a parameter. **(d)** A small P means the means are not all the same, but the distributions may still overlap quite a bit. (See the "Caution" immediately preceding this exercise in the text.)

12.7. Assuming the t (ANOVA) test establishes that the means are different, contrasts and multiple comparison provide no further useful information. (With two means, there is only one comparison to make, and it has already been made by the t test.)

12.9. (a) With $I = 5$ groups and $N = 35$, we have df $I - 1 = 4$ and $N - I = 30$. In Table E, we see that $2.69 < F < 3.25$. **(b)** The sketch on the right shows the observed F value and the critical values from Table E. **(c)** $0.025 < P < 0.050$ (software gives 0.0420). **(d)** The alternative hypothesis states that at least one mean is different, not that all means are different.

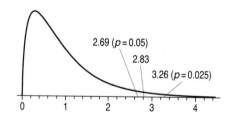

12.11. (a) $I = 3$ and $N = 33$, so the degrees of freedom are 2 and 30. $F = \frac{127}{50} = 2.54$. Comparing to the $F(2, 30)$ distribution in Table E, we find $2.49 < F < 3.32$, so $0.050 < P < 0.100$. (Software gives $P \doteq 0.0957$.) **(b)** $I = 4$ and $N = 32$, so the degrees of freedom are 3 and 28. $F = \frac{58/3}{182/28} \doteq 2.9744$. Comparing to the $F(3, 28)$ distribution in Table E, we find $2.95 < F < 3.63$, so $0.025 < P < 0.050$. (Software gives $P \doteq 0.0486$.)

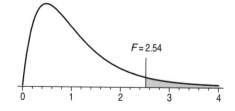 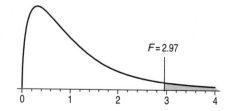

145

12.13. (a) Response: egg cholesterol level. Populations: chickens with different diets or drugs. $I = 3$, $n_1 = n_2 = n_3 = 25$, $N = 75$. **(b)** Response: rating on five-point scale. Populations: the three groups of students. $I = 3$, $n_1 = 31$, $n_2 = 18$, $n_3 = 45$, $N = 94$. **(c)** Response: quiz score. Populations: students in each TA group. $I = 3$, $n_1 = n_2 = n_3 = 14$, $N = 42$.

12.15. For all three situations, the hypotheses are H_0: $\mu_1 = \mu_2 = \mu_3$ versus H_a: at least one mean is different. The degrees of freedom are DFG = DFM = $I - 1$ ("model" or "between groups"), DFE = DFW = $N - I$ ("error" or "within groups"), and DFT = $N - 1$ ("total"). The degrees of freedom for the F test are DFG and DFE.

Situation	I	N	DFG	DFE	DFT	df for F statistic
Egg cholesterol level	3	75	2	72	74	$F(2, 72)$
Student opinions	3	94	2	91	93	$F(2, 91)$
Teaching assistants	3	42	2	39	41	$F(2, 39)$

12.17. (a) This sounds like a fairly well-designed experiment, so the results should at least apply to this farmer's breed of chicken. **(b)** It would be good to know what proportion of the total student body falls in each of these groups—that is, is anyone overrepresented in this sample? **(c)** How well a TA teaches one topic (power calculations) might not reflect that TA's overall effectiveness.

12.19. (a) With $I = 3$ and $N = 120$, we have df 2 and 117. **(b)** To use Table E, we compare to df 2 and 100; with $F > 5.02$, we conclude that $P < 0.001$. Software gives $P = 0.0003$. **(c)** Haggling and bargaining behavior is probably linked to the local culture, so we should hesitate to generalize these results beyond similar informal shops in Mexico.

12.21. (a) F can be made very small (close to 0), and P close to 1. **(b)** F increases, and P decreases. Moving the means farther apart means that (even with moderate spread) it is easier to see that the three groups represent three different populations (that is, populations having different means). Therefore, the evidence against H_0 becomes stronger.

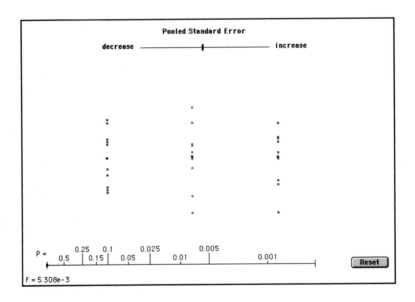

12.23. **(a)** Based on the sample means, fiber is cheapest and cable is most expensive. (Note that the providers are shown in this plot in the order given in the table, but they can be rearranged in any order.) **(b)** Yes; the smallest-to-largest standard deviation ratio is $\frac{40.39}{26.09} \doteq 1.55$. **(c)** The degrees of freedom are $I - 1 = 2$ and $N - I = 44$. From Table E (with df 2 and 40), we have $0.025 < P < 0.050$; software gives $P = 0.0427$. The difference in means is (barely) significant at the 5% level.

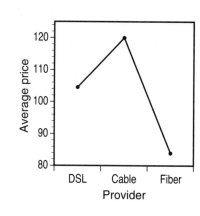

12.25. **(a)** Use matched pairs t methods; we examine the change in reaction time for each subject. **(b)** No: We cannot use ANOVA methods because we do not have four independent samples. (The same group of subjects performed each of the four tasks.)

12.27. **(a)** With $I = 4$ and $N = 2290$, the degrees of freedom are DFG $= I - 1 = 3$ and DFE $= N - I = 2286$. **(b)** MSE $= s_p^2 = 4.6656$, so $F = \frac{\text{MSG}}{\text{MSE}} = \frac{11.806}{4.6656} \doteq 2.5304$. **(c)** The $F(3, 1000)$ entry in Table E gives $0.05 < P < 0.10$; software give $P \doteq 0.0555$.

12.29. **(a)** The plot suggests that both drugs cause an increase in activity level, and Drug B appears to have a greater effect. **(b)** Yes: The guidelines for pooling standard deviations say that the ratio of largest to smallest should be less than 2; we have $\sqrt{\frac{14.00}{6.75}} \doteq 1.44 < 2$. The pooled variance is

$$s_p^2 = \frac{3(s_1^2 + s_2^2 + \cdots + s_5^2)}{3 + 3 + 3 + 3 + 3} = \frac{159}{15} = 10.6$$

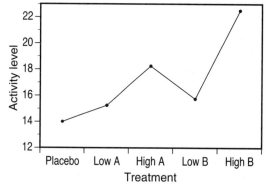

and $s_p = \sqrt{10.6} \doteq 3.2558$. **(c)** The degrees of freedom are DFG $= I - 1 = 4$ and DFE $= N - I = 15$. **(d)** Comparing to an $F(4, 15)$ distribution in Table E, we see that $3.80 < F < 4.89$, so $0.010 < P < 0.025$; software gives $P \doteq 0.0165$. We have significant evidence that the means are not all the same.

12.31. **(a)** The variation in sample size is some cause for concern, but there can be no extreme outliers in a 1-to-7 scale, so ANOVA is probably reliable. **(b)** Pooling is reasonable: $\frac{1.26}{1.03} \doteq 1.22 < 2$. **(c)** With $I = 5$ groups and total sample size $N = 410$, we use an $F(4, 405)$ distribution. We can compare 5.69 to an

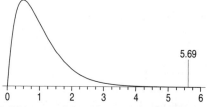

$F(4, 200)$ distribution in Table E and conclude that $P < 0.001$, or with software determine that $P \doteq 0.0002$. **(d)** Hispanic Americans have the highest emotion scores, Japanese are in the middle, and the other three cultures are the lowest (and very similar).

12.33. Because the descriptions of these contrasts do not specify an expected direction for the comparison, the subtraction could be done either way (in the order shown, or in the opposite order). **(a)** $\psi_1 = \mu_2 - \frac{1}{2}(\mu_1 + \mu_4)$. **(b)** $\psi_2 = \frac{1}{3}(\mu_1 + \mu_2 + \mu_4) - \mu_3$.

12.35. See the solution to Exercise 1.87 for stemplots. The means, standard deviations, and standard errors (all in millimeters) are given below. We reject H_0 and conclude that at least one mean is different ($F = 259.12$, df 2 and 51, $P < 0.0005$).

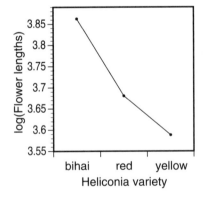

Variety	n	$\bar{x}$	s	$s_{\bar{x}}$
bihai	16	47.5975	1.2129	0.3032
red	23	39.7113	1.7988	0.3751
yellow	15	36.1800	0.9753	0.2518

Minitab output: Analysis of Variance on length

Source	DF	SS	MS	F	p
Factor	2	1082.87	541.44	259.12	0.000
Error	51	106.57	2.09		
Total	53	1189.44			

12.37. Stemplots are shown on the right; means, standard deviations, and standard errors are given below. We reject H_0 and conclude that at least one mean is different ($F = 244.27$, df 2 and 51, $P < 0.0005$). These results are essentially the same as in Exercise 12.35.

bihai		red		yellow	
38	33	36	233333	35	44
38	44444555	36	4445	35	667
38	7777	36	667	35	8889
38		36	8	36	00011
39	11	37	00	36	
		37	23333	36	4
		37	4		
		37	6		

   Note: *All of the numbers in these samples are between 34 and 50; for that range of input values, the log function closely resembles a line. Because a linear transformation has no effect on skewness, the effect of this transformation is minimal, as can be confirmed by comparing these stemplots to those in the solution to Exercise 1.87, and this means plot with the one in the solution to Exercise 12.35.*

Variety	n	$\bar{x}$	s	$s_{\bar{x}}$
bihai	16	3.8625	0.02515	0.006286
red	23	3.6807	0.04496	0.009374
yellow	15	3.5882	0.02698	0.006966

Minitab output: Analysis of Variance on log-length

Source	DF	SS	MS	F	p
Factor	2	0.61438	0.30719	244.27	0.000
Error	51	0.06414	0.00126		
Total	53	0.67852			

12.39. (a) The means, standard deviations, and standard errors are given on the following page (all in grams per cm^2). **(b)** All three distributions appear to reasonably close to Normal, and the standard deviations are suitable for pooling. **(c)** ANOVA gives $F = 7.72$ (df 2 and 42) and $P = 0.001$, so we conclude that the means are not all the same. **(d)** With df $= 42$, 3 comparisons, and $\alpha = 0.05$, the Bonferroni critical value is $t^{**} = 2.4937$. The pooled standard deviation is $s_p \doteq 0.01437$ and the standard error of each difference is $\mathrm{SE}_D = s_p\sqrt{1/15 + 1/15} \doteq 0.005246$, so two means are significantly different if they differ by $t^{**}\mathrm{SE}_D \doteq 0.01308$. The high-dose mean is significantly different from the other two. **(e)** Briefly: High doses of kudzu isoflavones increase BMD.

	n	$\bar{x}$	s	$s_{\bar{x}}$
Control	15	0.2189	0.01159	0.002992
Low dose	15	0.2159	0.01151	0.002972
High dose	15	0.2351	0.01877	0.004847

Minitab output: Analysis of Variance on BMD

Source	DF	SS	MS	F	p
Factor	2	0.003186	0.001593	7.72	0.001
Error	42	0.008668	0.000206		
Total	44	0.011853			

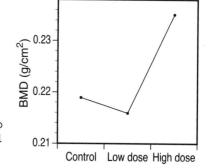

12.41. **(a)** The mean responses were not significantly different for the last question. **(b)** Taking the square roots of the given values of MSE gives the values of s_p. For the Bonferroni method with $\alpha = 0.05$, df $= 310$, and 10 comparisons, $t^{**} = 2.827$. Only the largest difference within each set of means is significant:

$$t_{14} = \frac{3.97 - 5.33}{1.8058\sqrt{\frac{1}{34} + \frac{1}{26}}} \doteq -2.891 \quad \text{experience culture—honeymoon and leisure groups}$$

$$t_{23} = \frac{3.38 - 2.39}{1.6855\sqrt{\frac{1}{56} + \frac{1}{105}}} \doteq 3.550 \quad \text{group tour—fraternal association/sports groups}$$

$$t_{45} = \frac{5.33 - 4.02}{2.0700\sqrt{\frac{1}{26} + \frac{1}{94}}} \doteq 2.856 \quad \text{ocean sports—leisure/business groups}$$

12.43. We have six comparisons to make, and df $= 74$, so the Bonferroni critical value with $\alpha = 0.05$ is $t^{**} = 2.7111$. The pooled standard deviation is $s_p \doteq 2.7348$. The table below shows the differences, their standard errors, and the t statistics.

The Piano mean is significantly higher than the other three, but the other three means are not significantly different.

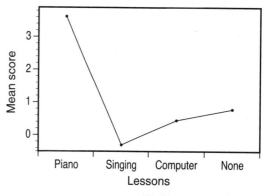

$D_{PS} =$	3.91765	$D_{PC} =$	3.16765	$D_{PN} =$	2.83193
$\mathrm{SE}_{PS} =$	0.98380	$\mathrm{SE}_{PC} =$	0.77066	$\mathrm{SE}_{PN} =$	0.86843
$t_{PS} =$	3.982	$t_{PC} =$	4.110	$t_{PN} =$	3.261
		$D_{SC} =$	-0.75000	$D_{SN} =$	-1.08571
		$\mathrm{SE}_{SC} =$	1.05917	$\mathrm{SE}_{SN} =$	1.13230
		$t_{SC} =$	-0.708	$t_{SN} =$	-0.959
				$D_{CN} =$	-0.33571
				$\mathrm{SE}_{CN} =$	0.95297
				$t_{CN} =$	-0.352

12.45. (a) Pooling is reasonable: The ratio is $\frac{0.824}{0.657} \doteq$ 1.25. For the pooled standard deviation, we compute

$$s_p^2 = \frac{488s_1^2 + 68s_2^2 + 211s_3^2}{488 + 68 + 211} \doteq 0.5902$$

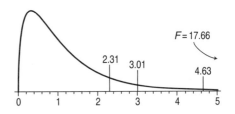

so $s_p \doteq \sqrt{0.5902} \doteq 0.7683$. **(b)** Comparing $F = 17.66$ to an $F(2, 767)$ distribution, we find $P < 0.001$. Sketches of this distribution will vary; in the graph on the right, the three marked points are the 10%, 5%, and 1% critical values, so we can see that the observed value lies well above the bulk of this distribution. **(c)** For the contrast $\psi = \mu_2 - \frac{1}{2}(\mu_1 + \mu_3)$, we test $H_0: \psi = 0$ versus $H_a: \psi > 0$. We find $c \doteq 0.585$ with $SE_c \doteq 0.0977$, so $t = c/SE_c \doteq 5.99$ with df $= 767$, and $P < 0.0001$.

12.47. (a) Pooling is reasonable, as the largest-to-smallest ratio is about 1.65. **(b)** ANOVA gives $F = 7.98$ (df 2 and 27), for which $P = 0.002$. We reject H_0 and conclude that not all means are equal.

	n	$\bar{x}$	s
Control	10	601.1	27.364
Low jump	10	612.5	19.329
High jump	10	638.7	16.594

Minitab output: Analysis of Variance on density

Source	DF	SS	MS	F	p
Treatment	2	7434	3717	7.98	0.002
Error	27	12580	466		
Total	29	20013			

12.49. (a) Pooling is risky because $\frac{0.6283}{0.2520} = 2.49 > 2$. **(b)** ANOVA gives $F = 31.16$ (df 2 and 9), for which $P < 0.0005$. We reject H_0 and conclude that not all means are equal.

	n	$\bar{x}$	s
Aluminum	4	2.0575	0.2520
Clay	4	2.1775	0.6213
Iron	4	4.6800	0.6283

Minitab output: Analysis of Variance on iron

Source	DF	SS	MS	F	p
Pot	2	17.539	8.770	31.16	0.000
Error	9	2.533	0.281		
Total	11	20.072			

12.51. (a) Pooling is risky because $\frac{8.66}{2.89} = 3 > 2$. **(b)** ANOVA gives $F = 137.94$ (df 5 and 12), for which $P < 0.0005$. We reject H_0 and conclude that not all means are equal.

	n	$\bar{x}$	s
ECM1	3	65.0%	8.6603%
ECM2	3	63.$\overline{3}$%	2.8868%
ECM3	3	73.$\overline{3}$%	2.8868%
MAT1	3	23.$\overline{3}$%	2.8868%
MAT2	3	6.$\overline{6}$%	2.8868%
MAT3	3	11.$\overline{6}$%	2.8868%

Minitab output: Analysis of Variance on Gpi

Source	DF	SS	MS	F	p
Treatment	5	13411.1	2682.2	137.94	0.000
Error	12	233.3	19.4		
Total	17	13644.4			

12.53. Let μ_1 be the placebo mean, μ_2 and μ_3 be the means for low and high doses of Drug A, and μ_4 and μ_5 be the means for low and high doses of Drug B. Recall that $s_p \doteq 3.2558$. **(a)** The first contrast is $\psi_1 = \mu_1 - \frac{1}{2}(\mu_2 + \mu_4)$; the second is $\psi_2 = \mu_3 - \mu_2 - (\mu_5 - \mu_4)$. **(b)** The estimated contrasts are $c_1 = 14.00 - 0.5(15.25) - 0.5(15.75) = -1.5$ and

$c_2 = (18.25 - 15.25) - (22.50 - 15.75) = -3.75$. The respective standard errors are:

$$SE_{c_1} = s_p \sqrt{\frac{1}{4} + \frac{0.25}{4} + \frac{0}{4} + \frac{0.25}{4} + \frac{0}{4}} \doteq 1.9937 \text{ and}$$

$$SE_{c_2} = s_p \sqrt{\frac{0}{4} + \frac{1}{4} + \frac{1}{4} + \frac{1}{4} + \frac{1}{4}} = s_p \doteq 3.2558$$

(c) Neither contrast is significant ($t_1 \doteq -0.752$ and $t_2 \doteq -1.152$, for which the one-sided P-values are 0.2317 and 0.1337). We do not have enough evidence to conclude that low doses increase activity level over a placebo, nor can we conclude that activity level changes due to increased dosage are different between the two drugs.

12.55. (a) The plot (below) shows granularity (which varies between groups), but that should not make us question independence; it is due to the fact that the scores are all integers. **(b)** The ratio of the largest to the smallest standard deviations is $1.595/0.931 \doteq 1.714$—less than 2. **(c)** Apart from the granularity, the quantile plots (below and on the following page) are reasonably straight. **(d)** Again, apart from the granularity, the residual quantile plot (below, right) looks pretty good.

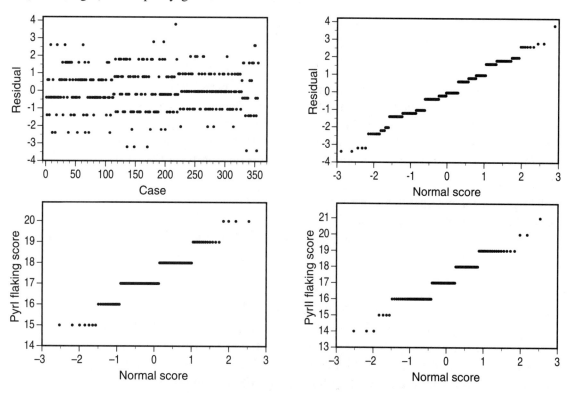

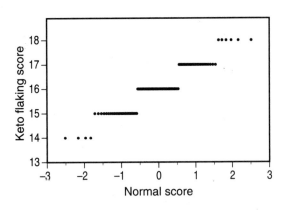

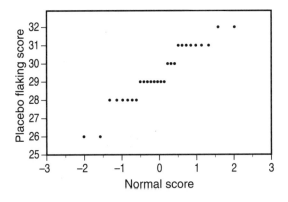

12.57. (a) The three contrasts are:

$$\psi_1 = \tfrac{1}{3}\mu_{Py1} + \tfrac{1}{3}\mu_{Py2} + \tfrac{1}{3}\mu_K - \mu_P$$
$$\psi_2 = \tfrac{1}{2}\mu_{Py1} + \tfrac{1}{2}\mu_{Py2} - \mu_K$$
$$\psi_3 = \mu_{Py1} - \mu_{Py2}$$

$c_1 = -12.51$	$c_2 = 1.269$	$c_3 = 0.191$
$SE_{c_1} \doteq 0.2355$	$SE_{c_2} \doteq 0.1413$	$SE_{c_3} \doteq 0.1609$
$t_1 = -53.17$	$t_2 = 8.98$	$t_3 = 1.19$
$P_1 < 0.0005$	$P_2 < 0.0005$	$P_3 \doteq 0.2359$

(b) The pooled standard deviation is $s_p = \sqrt{MSE} \doteq 1.1958$. The estimated contrasts and their standard errors are in the table. For example:

$$SE_{c_1} = s_p\sqrt{\tfrac{1}{9}/112 + \tfrac{1}{9}/109 + \tfrac{1}{9}/106 + 1/28} \doteq 0.2355$$

(c) We test H_0: $\psi_i = 0$ versus H_a: $\psi_i \neq 0$ for each contrast. The t- and P-values are given in the table. The Placebo mean is significantly higher than the average of the other three, while the Keto mean is significantly lower than the average of the two Pyr means. The difference between the Pyr means is not significant (meaning the second application of the shampoo is of little benefit)—this agrees with our conclusion from Exercise 12.56.

12.59. Because the measurements in Exercise 12.51 are percents, the instructions to "add 5% to each response" could be interpreted in two ways:

(1) new response = old response + 5
(2) new response = old response × 1.05

The table on the right gives summary statistics for both interpretations (all numbers in per-

	Version (1)		Version (2)	
	$\bar{x}$	s	$\bar{x}$	s
ECM1	70.0	8.6603	68.25	9.0933
ECM2	68.$\overline{3}$	2.8868	66.5	3.0311
ECM3	78.$\overline{3}$	2.8868	77	3.0311
MAT1	28.$\overline{3}$	2.8868	24.5	3.0311
MAT2	11.$\overline{6}$	2.8868	7	3.0311
MAT3	16.$\overline{6}$	2.8868	12.25	3.0311

cents). For (1), the means increase by 5, but everything else remains the same; the ANOVA table is identical to the one in the solution to Exercise 12.51. For (2), both the means and standard deviations are multiplied by 1.05, SS and MS are multiplied by 1.05^2, but F and P remain the same (ANOVA table below).

Minitab output: Analysis of Variance on GpiVers2

```
Source      DF      SS        MS        F        p
Treatment    5   14785.8    2957.1   137.94    0.000
Error       12     257.3      21.4
Total       17   15043.0
```

12.61. A table of means and standard deviations is below. Quantile plots are not shown, but apart from the granularity of the scores and a few possible outliers, there are no marked deviations from Normality. Pooling is reasonable for both PRE1 and PRE2; the ratios are 1.24 and 1.48.

For both PRE1 and PRE2, we test H_0: $\mu_B = \mu_D = \mu_S$ versus H_a: at least one mean is different. Both tests have df 2 and 63. For PRE1, $F = 1.13$ and $P = 0.329$; for PRE2, $F = 0.11$ and $P = 0.895$. There is no reason to believe that the mean pretest scores differ between methods.

		PRE1		PRE2	
Method	n	$\bar{x}$	s	$\bar{x}$	s
Basal	22	10.5	2.9721	$5.\overline{27}$	2.7634
DRTA	22	$9.\overline{72}$	2.6936	$5.\overline{09}$	1.9978
Strat	22	$9.1\overline{36}$	3.3423	$4.9\overline{54}$	1.8639

Minitab output: Analysis of Variance on PRE1

Source	DF	SS	MS	F	p
Group	2	20.58	10.29	1.13	0.329
Error	63	572.45	9.09		
Total	65	593.03			

Analysis of Variance on PRE2

Source	DF	SS	MS	F	p
Group	2	1.12	0.56	0.11	0.895
Error	63	317.14	5.03		
Total	65	318.26			

12.63. The scatterplot (below, left) suggests that a straight line is *not* the best choice of a model. Regression gives the formula

$$\widehat{\text{Score}} = 4.432 - 0.000102 \text{ Friends}$$

Not surprisingly, the slope is not significantly different from 0 ($t = -0.28$, $P = 0.782$). The regression only explains 0.1% of the variation in score. The residual plot (below, right) is nearly identical to the first scatterplot, and suggests (as that did) that a quadratic model might be a better choice.

Note: *If one fits a quadratic model, it does better (and has significant coefficients), but it still only explains 8.3% of the variation in attractiveness.*

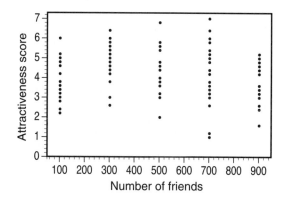
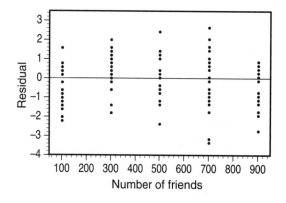

Minitab output: Regression of attractiveness score on number of friends

The regression equation is Score = 4.43 -0.000102 Friends

Predictor	Coef	Stdev	t-ratio	p
Constant	4.4321	0.2060	21.51	0.000
Friends	-0.0001023	0.0003694	-0.28	0.782

s = 1.150   R-sq = 0.1%   R-sq(adj) = 0.0%

12.67. (a) Sampling plans will vary but should attempt to address how cultural groups will be determined: Can we obtain such demographic information from the school administration? Do we simply select a large sample then poll each student to determine if he or she belongs to one of these groups? **(b)** Answers will vary with choice of H_a and desired power. For example, with the alternative $\mu_1 = \mu_2 = 4.4$, $\mu_3 = 5$, and standard deviation $\sigma = 1.2$, three samples of size 75 will produce power 0.89. (See G•Power output below.) **(c)** The report should make an attempt to explain the statistical issues involved; specifically, it should convey that sample sizes are sufficient to detect anticipated differences among the groups.

G•Power output

Post-hoc analysis for "F-Test (ANOVA)", Global, Groups: 3:
Alpha: 0.0500
Power (1-beta): 0.8920
Effect size "f": 0.2357
Total sample size: 225
Critical value: F(2,222) = 3.0365
Lambda: 12.4998

12.69. The design can be similar, although the types of music might be different. Bear in mind that spending at a casual restaurant will likely be less than at the restaurants examined in Exercise 12.28; this might also mean that the standard deviations could be smaller. A pilot study might be necessary to get an idea of the size of the standard deviations. Decide how big a difference in mean spending you would want to detect, then do some power computations.

Chapter 13 Solutions

13.1. (a) Two-way ANOVA is used when there are two factors (explanatory variables). (The outcome [response] variable is assumed to have a Normal distribution, meaning that it can take any value, at least in theory.) **(b)** Each level of A should occur with all three levels of B. (Level A has two factors.) **(c)** The RESIDUAL part of the model represents the error. **(d)** DFAB = $(I - 1)(J - 1)$.

13.3. (a) A *large* value of the AB F statistic indicates that we should reject the hypothesis of no interaction. **(b)** The relationship is backwards: *Mean* squares equal *sum of* squares divided by degrees of freedom. **(c)** Under H_0, the ANOVA test statistics have an F distribution. **(d)** If the sample sizes are not the same, the sums of squares may not add for "some methods of analysis." (See the 'Caution' on page 680; for more detail, see `http://afni.nimh.nih.gov/sscc/gangc/SS.html`.)

13.5. A 3×2 ANOVA with 5 observations per cell has $I = 3$, $J = 2$, and $N = 30$. **(a)** The degrees of freedom for interaction are DFAB = $(I - 1)(J - 1) = 2$ and DFE = $N - IJ = 24$. The five critical values from Table E are 2.54, 3.40, 4.32, 5.61, and 9.34. **(b)** The sketch on the right shows the observed F-

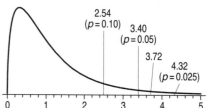

value given in part (c) and the bounding critical values from Table E. **(c)** In Table E, we see that $3.40 < F < 4.32$, so $0.025 < P < 0.05$. (Software gives 0.0392.) **(d)** The mean profiles would not look parallel because the interaction term is significantly different from 0.

13.7. (a) The factors are gender ($I = 2$) and age ($J = 3$). The response variable is the percent of pretend play. The total number of observations is $N = (2)(3)(11) = 66$. **(b)** The factors are time after harvest ($I = 5$) and amount of water ($J = 2$). The response variable is the percent of seeds germinating. The total number of observations is $N = 30$ (3 lots of seeds in each of the 10 treatment combinations). **(c)** The factors are mixture ($I = 6$) and freezing/thawing cycles ($J = 3$). The response variable is the strength of the specimen. The total number of observations is $N = 54$. **(d)** The factors are training programs ($I = 4$) and the number of days to give the training ($J = 2$). The response variable is not specified, but presumably is some measure of the training's effectiveness. The total sample size is $N = 80$.

13.9. **(a)** There appears to be an interaction: A thank-you note increases repurchase intent by over 1 point for those with short history, and decreases it (very slightly) for customers with long history. Note that either variable could be on the horizontal axis in the plot of means. **(b)** The marginal means are

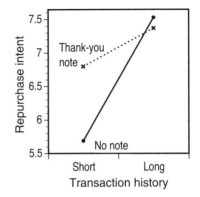

| Short history | 6.245 | No thank-you note | 6.61 |
| Long history | 7.45 | Thank-you note | 7.085 |

For example, $\frac{5.69+6.80}{2} = 6.245$. The history marginal means convey the fact that repurchase intent is higher for customers with long history. The thank-you note marginal means suggest that a thank-you note increases repurchase intent, but they are harder to interpret because of the interaction.

13.11. **(a)** The plot suggests a possible interaction because the means are not parallel. (Note that we could have chosen to put dish type on the horizontal axis instead of proximity; either explanatory variable will do.) **(b)** By subjecting the same individual to all four treatments, rather than four individuals to one treatment each, we reduce the within-groups variability (the residual), which makes it easier to detect between-groups variability (the main effects and interactions).

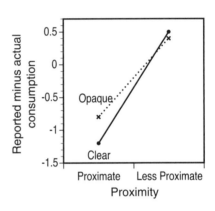

13.13. **(a)** There may be an interaction: For a favorable process, a favorable outcome increases satisfaction quite a bit more than for an unfavorable process (+2.32 versus +0.24). **(b)** With humor, the increase in satisfaction from a favorable outcome is less for a favorable process (+0.49 compared to +1.32). **(c)** There seems to be a three-factor interaction, because the interactions in parts (a) and (b) are different.

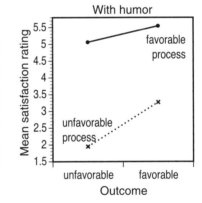

13.15. Marginal means are listed in the table on the following page. In each case, we find the average of the four means for each level of the characteristic. For example, for Humor, we have

$$\text{No humor:} \quad \frac{3.04 + 5.36 + 2.84 + 3.08}{4} = 3.58$$

$$\text{Humor:} \quad \frac{5.06 + 5.55 + 1.95 + 3.27}{4} = 3.9575$$

The presence of humor slightly increases mean satisfaction. The process and outcome effects appear to be greater (that is, the change in mean satisfaction is greater).

Marginal means

Humor		Process		Outcome	
No	3.58	Favorable	4.7525	Favorable	4.315
Yes	3.9575	Unfavorable	2.785	Unfavorable	3.2225

13.17. For the pooled standard deviation, we first find

$$s_p^2 = \frac{(237)(1.668^2) + (124)(1.909^2) + \cdots + (86)(1.875^2)}{237 + 124 + \cdots + 86} \doteq \frac{2535.19}{805} \doteq 3.1493$$

so $s_p \doteq \sqrt{3.1493} \doteq 1.7746$. The largest-to-smallest standard deviation ratio is $\frac{2.024}{1.601} \doteq 1.26 < 2$, so it is reasonable to use this pooled estimate.

13.19. Means plots are below. Possible observations: Except for female responses to purchase intention, means decreased from Canada to the United States to France. Females had higher means than men in almost every case, except for French responses to credibility and purchase intention (suggesting a modest interaction). Gender differences in France are considerably smaller than in either Canada or the United States.

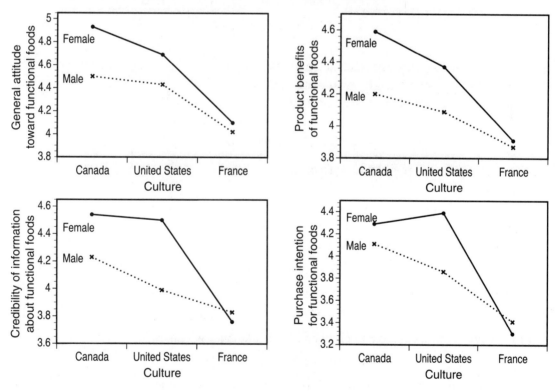

13.21. (a) The marginal means (as well as the individual cell means) are in the table below. The first two means suggest that the intervention group showed more improvement than the control group. **(b)** Interaction means that the mean number of actions changes differently over time for the two groups. We see this in the plot below because the lines connecting the means are not parallel.

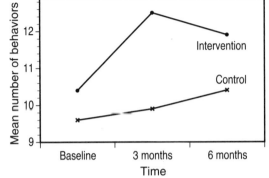

	Time			
Group	Baseline	3 mo.	6 mo.	Mean
Intervention	10.4	12.5	11.9	11.6
Control	9.6	9.9	10.4	9.967
Mean	10.0	11.2	11.15	10.783

13.23. We have $I = 3$, $J = 2$, and $N = 30$, so the degrees of freedom are DFA $= 2$, DFB $= 1$, DFAB $= 2$, and DFE $= 24$. This allows us to determine P-values (or to compare to Table E), and we find that there are no significant effects (although B is close):

$$F_A = 1.87 \text{ has df 2 and 24, so } P = 0.1759$$

$$F_B = 3.49 \text{ has df 1 and 24, so } P = 0.0740$$

$$F_{AB} = 2.14 \text{ has df 2 and 24, so } P = 0.1396$$

13.25. (a) The means are nearly parallel, and show little evidence of an interaction. **(b)** With equal sample sizes, the pooled variance is simply the unweighted average of the variances:

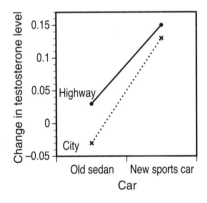

$$s_p^2 = \frac{1}{4}(0.12^4 + 0.14^2 + 0.12^2 + 0.13^2) = 0.016325$$

Therefore, $sp = \sqrt{0.016325} \doteq 0.1278$. **(c)** Note that all of these contrasts have been arranged so that, if the researchers' suspicions are correct, the contrast will be positive. To compare new-car testosterone change to old-car change, the appropriate contrast is

$$\psi_1 = \tfrac{1}{2}(\mu_{\text{new,city}} + \mu_{\text{new,highway}}) - \tfrac{1}{2}(\mu_{\text{old,city}} + \mu_{\text{old,highway}})$$

To compare city change to highway change for new cars, we take

$$\psi_2 = \mu_{\text{new,city}} - \mu_{\text{new,highway}}$$

To compare highway change to city change for old cars, we take

$$\psi_3 = \mu_{\text{old,highway}} - \mu_{\text{old,city}}$$

(d) By subjecting the same individual to all four treatments, rather than four individuals to one treatment each, we reduce the within-groups variability (the residual), which makes it easier to detect between-groups variability (the main effects and interactions).

13.27. (a) Plot on the right. (b) There seems to be a fairly large difference between the means based on how much the rats were allowed to eat, but not very much difference based on the chromium level. There may be an interaction: the NM mean is lower than the LM mean, while the NR mean is higher than the LR mean. (c) The marginal means are L: 4.86, N: 4.871, M: 4.485, R: 5.246. For low chromium level (L), R minus M is 0.63; for normal chromium (N), R minus M is 0.892. Mean GITH levels are lower for M than for R; there is not much difference for L versus N. The difference between M and R is greater among rats who had normal chromium levels in their diets (N).

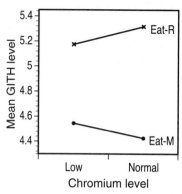

13.29. The "Other" category had the lowest mean SATM score for both genders; this is apparent from the graph (right) and from the marginal means (CS: 605, EO: 624.5, O: 566.) Males had higher mean scores in CS and O, while females are slightly higher in EO; this indicates an interaction. Overall, the marginal means by gender are 611.7 (males) and 585.3 (females).

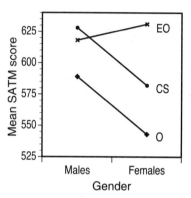

13.31. (a) The pooled variance is

$$s_p^2 = \frac{(31)(36.4^2) + (24)(31.2^2) + (24)(41.6^2) + (26)(42.4^2)}{31 + 24 + 24 + 26}$$

$$= \frac{152,711.52}{105} \doteq 1454.4$$

so $s_p \doteq \sqrt{1454.4} \doteq 38.14$. There were $N = 109$ items in the sample, and four groups, so df $= 105$. (b) Pooling is reasonable because the ratio of the largest and smallest standard deviations is $\frac{42.4}{31.2} \doteq 1.36 < 2$. (c) The marginal means are:

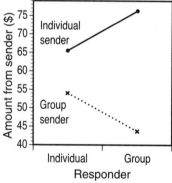

Sender		Responder	
Individual:	$\frac{1}{2}(65.5 + 76.3) = \$70.90$	Individual:	$\frac{1}{2}(65.5 + 54.0) = \$59.75$
Group:	$\frac{1}{2}(54.0 + 43.7) = \$48.85$	Group:	$\frac{1}{2}(76.3 + 43.7) = \$60.00$

(d) There appears to be an interaction: Individuals send more money to groups, while groups send more money to individuals. (e) Compare the statistics to an $F(1, 105)$ distribution. The three P-values are 0.0033 (sender), 0.9748 (responder), and 0.1522 (interaction). Only the main effect of sender is significant.

13.33. Yes; the iron-pot means are the highest, and the F statistic for testing the effect of the pot type is very large. (In this case, the interaction does not weaken any evidence that iron-pot foods contain more iron; it only suggests that while iron pots increase iron levels in all foods, the effect is strongest for meats.)

13.35. (a) For all tool/time combinations, $n = 3$. Means and standard deviations are in the table on the following page. Note that five cells had no variability ($s = 0$). **(b)** Plot on the right. Except for tool 1, mean diameter is highest at time 2. Tool 1 had the highest mean diameters, followed by tool 2, tool 4, tool 3, and tool 5. **(c)** Minitab output below; all F statistics are highly significant. **(d)** There is strong

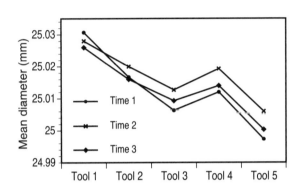

evidence of a difference in mean diameter among the tools and among the times. There is also an interaction (specifically, tool 1's mean diameters changed differently over time compared to the other tools).

Minitab output: Two-way ANOVA for diameter on tool and time

Source	DF	SS	MS	F	P
Tool	4	0.00359720	0.00089930	412.94	0.000
Time	2	0.00018991	0.00009496	43.60	0.000
Tool*Time	8	0.00013320	0.00001665	7.65	0.000
Error	30	0.00006533	0.00000218		
Total	44	0.00398564			

Diameter (mm)

Tool	Time 1 (8:00AM) $\bar{x}$	s	Time 2 (11:00AM) $\bar{x}$	s	Time 3 (3:00PM) $\bar{x}$	s
1	25.0307	0.001155	25.0280	0	25.0260	0
2	25.0167	0.001155	25.0200	0.002000	25.0160	0
3	25.0063	0.001528	25.0127	0.001155	25.0093	0.001155
4	25.0120	0	25.0193	0.001155	25.0140	0.004000
5	24.9973	0.001155	25.0060	0	25.0003	0.001528

13.37. (a) All three F values have df 1 and 945, and the P-values are < 0.001, < 0.001, and 0.1477. Gender and handedness both have significant effects on mean lifetime, but there is no significant interaction. **(b)** Women live about 6 years longer than men (on the average), while right-handed people average 9 more years of life than left-handed people. "There is no interaction" means that handedness affects both genders in the same way, and vice versa.

13.39. (a) & **(b)** The table below lists the means and standard deviations (the latter in parentheses) of the nitrogen contents of the plants. The two plots below suggest that plant 1 and plant 3 have the highest nitrogen content, plant 2 is in the middle, and plant 4 is the lowest. (In the second plot, the points are so crowded together that no attempt was made to differentiate among the different water levels.) There is no consistent effect of water level on nitrogen content. Standard deviations range from 0.0666 to 0.3437, for a ratio of 5.16—larger than we like. **(c)** Minitab output below. Both main effects and the interaction are highly significant.

	Amount of water per day						
Species	50mm	150mm	250mm	350mm	450mm	550mm	650mm
1	3.2543	2.7636	2.8429	2.9362	3.0519	3.0963	3.3334
	(0.2287)	(0.0666)	(0.2333)	(0.0709)	(0.0909)	(0.0815)	(0.2482)
2	2.4216	2.0502	2.0524	1.9673	1.9560	1.9839	2.2184
	(0.1654)	(0.1454)	(0.1481)	(0.2203)	(0.1571)	(0.2895)	(0.1238)
3	3.0589	3.1541	3.2003	3.1419	3.3956	3.4961	3.5437
	(0.1525)	(0.3324)	(0.2341)	(0.2965)	(0.2533)	(0.3437)	(0.3116)
4	1.4230	1.3037	1.1253	1.0087	1.2584	1.2712	0.9788
	(0.1738)	(0.2661)	(0.1230)	(0.1310)	(0.2489)	(0.0795)	(0.2090)

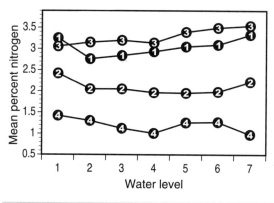 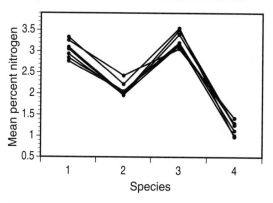

Minitab output: Two-way ANOVA for Pctnit on species and water

```
Source          DF       SS        MS        F       P
Species          3   172.3916   57.4639  1301.32  0.000
Water            6     2.5866    0.4311     9.76  0.000
Species*Water   18     4.7446    0.2636     5.97  0.000
Error          224     9.8914    0.0442
Total          251   189.6143
```

13.41. For each water level, there is highly significant evidence of variation in nitrogen level among plant species (Minitab output below). For each water level, we have df = 32, 6 comparisons, and $\alpha = 0.05$, so the Bonferroni critical value is $t^{**} = 2.8123$. (If we take into account that there are 7 water levels, so

Water level	s_p	SE_D	MSD1	MSD2
1	0.1824	0.0860	0.2418	0.3059
2	0.2274	0.1072	0.3015	0.3814
3	0.1912	0.0902	0.2535	0.3208
4	0.1991	0.0939	0.2640	0.3340
5	0.1994	0.0940	0.2643	0.3344
6	0.2318	0.1093	0.3073	0.3887
7	0.2333	0.1100	0.3093	0.3913

that overall we are performing $6 \times 7 = 42$ comparisons, we should take $t^{**} = 3.5579$.) The table on the right gives the pooled standard deviations s_p, the standard errors of each difference $SE_D = s_p\sqrt{1/9 + 1/9}$, and the "minimum significant difference" $MSD = t^{**}SE_D$ (two means are significantly different if they differ by at least this amount). MSD1 uses $t^{**} = 2.8123$, and MSD2 uses $t^{**} = 3.5579$. As it happens, for either choice of MSD, the only *non*significant differences are between species 1 and 3 for water levels 1, 4, and 7. (These are the three closest pairs of points in the plot from the solution to Exercise 13.39.) Therefore, for every water level, species 4 has the lowest nitrogen level and species 2 is next. For water levels 1, 4, and 7, species 1 and 3 are statistically tied for the highest level; for the other levels, species 3 is the highest, with species 1 coming in second.

Minitab output: One-way ANOVA on species for water level 1

Source	DF	SS	MS	F	p
Species	3	18.3711	6.1237	184.05	0.000
Error	32	1.0647	0.0333		
Total	35	19.4358			

One-way ANOVA on species for water level 2

Source	DF	SS	MS	F	p
Species	3	17.9836	5.9945	115.93	0.000
Error	32	1.6546	0.0517		
Total	35	19.6382			

One-way ANOVA on species for water level 3

Source	DF	SS	MS	F	p
Species	3	22.9171	7.6390	208.87	0.000
Error	32	1.1704	0.0366		
Total	35	24.0875			

One-way ANOVA on species for water level 4

Source	DF	SS	MS	F	p
Species	3	25.9780	8.6593	218.37	0.000
Error	32	1.2689	0.0397		
Total	35	27.2469			

One-way ANOVA on species for water level 5

Source	DF	SS	MS	F	p
Species	3	26.2388	8.7463	220.01	0.000
Error	32	1.2721	0.0398		
Total	35	27.5109			

One-way ANOVA on species for water level 6

Source	DF	SS	MS	F	p
Species	3	28.0648	9.3549	174.14	0.000
Error	32	1.7191	0.0537		
Total	35	29.7838			

One-way ANOVA on species for water level 7

Source	DF	SS	MS	F	p
Species	3	37.5829	12.5276	230.17	0.000
Error	32	1.7417	0.0544		
Total	35	39.3246			

13.43. (a) & (b) The tables below list the means and standard deviations (the latter in parentheses). The means plots show that biomass (both fresh and dry) increases with water level for all plants. Generally, plants 1 and 2 have higher biomass for each water level, while plants 3 and 4 are lower. Standard deviation ratios are quite high for both fresh and dry biomass: $108.01/6.79 \doteq 15.9$ and $35.76/3.12 \doteq 11.5$. **(c)** Minitab output on the following page. For both fresh and dry biomass, main effects and the interaction are significant. (The interaction for fresh biomass has $P = 0.04$; other P-values are smaller.)

Fresh biomass

Species	50mm	150mm	250mm	350mm	450mm	550mm	650mm
1	109.095	165.138	168.825	215.133	258.900	321.875	300.880
	(20.949)	(29.084)	(18.866)	(42.687)	(45.292)	(46.727)	(29.896)
2	116.398	156.750	254.875	265.995	347.628	343.263	397.365
	(29.250)	(46.922)	(13.944)	(59.686)	(54.416)	(98.553)	(108.011)
3	55.600	78.858	90.300	166.785	164.425	198.910	188.138
	(13.197)	(29.458)	(28.280)	(41.079)	(18.646)	(33.358)	(18.070)
4	35.128	58.325	94.543	96.740	153.648	175.360	158.048
	(11.626)	(6.789)	(13.932)	(24.477)	(22.028)	(32.873)	(70.105)

Dry biomass

Species	50mm	150mm	250mm	350mm	450mm	550mm	650mm
1	40.565	63.863	71.003	85.280	103.850	136.615	120.860
	(5.581)	(7.508)	(6.032)	(10.868)	(15.715)	(16.203)	(17.137)
2	34.495	57.365	79.603	95.098	106.813	103.180	119.625
	(11.612)	(6.149)	(13.094)	(25.198)	(18.347)	(25.606)	(35.764)
3	26.245	31.865	36.238	64.800	64.740	74.285	67.258
	(6.430)	(11.322)	(11.268)	(9.010)	(3.122)	(12.277)	(7.076)
4	15.530	23.290	37.050	34.390	48.538	61.195	53.600
	(4.887)	(3.329)	(5.194)	(11.667)	(5.658)	(12.084)	(25.290)

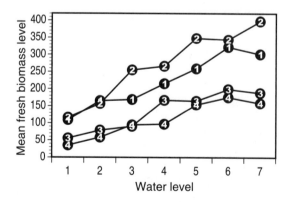 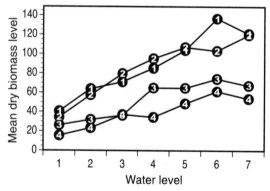

Minitab output: Two-way ANOVA for fresh biomass

Source	DF	SS	MS	F	P
Species	3	458295	152765	81.45	0.000
Water	6	491948	81991	43.71	0.000
Species*Water	18	60334	3352	1.79	0.040
Error	84	157551	1876		
Total	111	1168129			

Two-way ANOVA for dry biomass

Source	DF	SS	MS	F	P
Species	3	50523.8	16841.3	79.93	0.000
Water	6	56623.6	9437.3	44.79	0.000
Species*Water	18	8418.8	467.7	2.22	0.008
Error	84	17698.4	210.7		
Total	111	133264.6			

13.45. For each water level, there is highly significant evidence of variation in biomass level (both fresh and dry) among plant species (Minitab output follows). For each water level, we have df = 12, 6 comparisons, and $\alpha = 0.05$, so the Bonferroni critical value is $t^{**} = 3.1527$. (If we take into account that there are 7 water levels, so that overall we are performing $6 \times 7 = 42$ comparisons, we should take $t^{**} = 4.2192$.) The table below gives the pooled standard deviations s_p, the standard errors of each difference $SE_D = s_p\sqrt{1/4 + 1/4}$, and the "minimum significant difference" $MSD = t^{**}SE_D$ (two means are significantly different if they differ by at least this amount). MSD1 uses $t^{**} = 3.1527$, and MSD2 uses $t^{**} = 4.2192$. Rather than give a full listing of which differences are significant, we note that plants 3 and 4 are *not* significantly different, nor are 1 and 3 (except for one or two water levels). All other plant combinations are significantly different for at least three water levels. For fresh biomass, plants 2 and 4 are different for *all* levels, and for dry biomass, 1 and 4 differ for all levels.

	Fresh biomass				Dry biomass			
Water level	s_p	SE_D	MSD1	MSD2	s_p	SE_D	MSD1	MSD2
1	20.0236	14.1588	44.6382	50.3764	7.6028	5.3760	16.9487	19.1274
2	31.4699	22.2526	70.1552	79.1735	7.6395	5.4019	17.0305	19.2197
3	19.6482	13.8934	43.8012	49.4318	9.5103	6.7248	21.2010	23.9263
4	43.7929	30.9663	97.6265	110.1762	15.5751	11.0133	34.7213	39.1846
5	38.2275	27.0310	85.2197	96.1746	12.5034	8.8412	27.8734	31.4565
6	59.3497	41.9666	132.3068	149.3147	17.4280	12.3235	38.8518	43.8462
7	66.7111	47.1719	148.7174	167.8348	23.7824	16.8167	53.0176	59.8329

Minitab output: One-way ANOVA for fresh biomass — water level 1

Source	DF	SS	MS	F	p
Species	3	19107	6369	15.88	0.000
Error	12	4811	401		
Total	15	23918			

One-way ANOVA for fresh biomass — water level 2

Source	DF	SS	MS	F	p
Species	3	35100	11700	11.81	0.001
Error	12	11884	990		
Total	15	46984			

One-way ANOVA for fresh biomass — water level 3

Source	DF	SS	MS	F	p
Species	3	71898	23966	62.08	0.000
Error	12	4633	386		
Total	15	76531			

One-way ANOVA for fresh biomass — water level 4				
Source	DF	SS	MS	F	p
Species	3	62337	20779	10.83	0.001
Error	12	23014	1918		
Total	15	85351			

One-way ANOVA for fresh biomass — water level 5				
Source	DF	SS	MS	F	p
Species	3	99184	33061	22.62	0.000
Error	12	17536	1461		
Total	15	116720			

One-way ANOVA for fresh biomass — water level 6				
Source	DF	SS	MS	F	p
Species	3	86628	28876	8.20	0.003
Error	12	42269	3522		
Total	15	128897			

One-way ANOVA for fresh biomass — water level 7				
Source	DF	SS	MS	F	p
Species	3	144376	48125	10.81	0.001
Error	12	53404	4450		
Total	15	197780			

One-way ANOVA for dry biomass — water level 1				
Source	DF	SS	MS	F	p
Species	3	1411.2	470.4	8.14	0.003
Error	12	693.6	57.8		
Total	15	2104.8			

One-way ANOVA for dry biomass — water level 2				
Source	DF	SS	MS	F	p
Species	3	4597.1	1532.4	26.26	0.000
Error	12	700.3	58.4		
Total	15	5297.4			

One-way ANOVA for dry biomass — water level 3				
Source	DF	SS	MS	F	p
Species	3	6127.2	2042.4	22.58	0.000
Error	12	1085.3	90.4		
Total	15	7212.6			

One-way ANOVA for dry biomass — water level 4				
Source	DF	SS	MS	F	p
Species	3	8634	2878	11.86	0.001
Error	12	2911	243		
Total	15	11545			

One-way ANOVA for dry biomass — water level 5				
Source	DF	SS	MS	F	p
Species	3	10026	3342	21.38	0.000
Error	12	1876	156		
Total	15	11902			

One-way ANOVA for dry biomass — water level 6				
Source	DF	SS	MS	F	p
Species	3	13460	4487	14.77	0.000
Error	12	3645	304		
Total	15	17105			

One-way ANOVA for dry biomass — water level 7				
Source	DF	SS	MS	F	p
Species	3	14687	4896	8.66	0.002
Error	12	6787	566		
Total	15	21474			

13.47. (a) With $I = 2$, $J = 3$, and $N = 180$, the numerator degrees of freedom are $I - 1$, $J - 1$, and $(I - 1)(J - 1)$, respectively, and the denominator degrees of freedom for all three tests is DFE $= N - IJ = 174$:

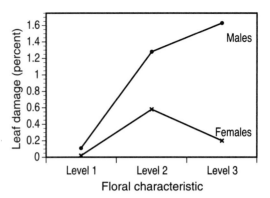

Source	df
Gender	1 and 174
Floral characteristic	2 and 174
Interaction	2 and 174

(b) Damage to males was higher for all characteristics. For males, damage was highest under characteristic level 3, while for females, the highest damage occurred at level 2. **(c)** Three of the standard deviations are at least half as large as the means. Because the response variable (leaf damage) had to be nonnegative, this suggests that these distributions are right-skewed; taking logarithms reduces the skewness.

13.49. The table and plot of the means suggest that females have higher HSE grades than males. For a given gender, there is not too much difference among majors. Normal quantile plots show no great deviations from Normality, apart from the granularity of the grades (most evident among women in EO). In the ANOVA, only the effect of gender is significant. Residual analysis (not shown) reveals some causes for concern; for example, the variance does not appear to be constant.

Minitab output: Two-way ANOVA for HSE on sex and major

Source	DF	SS	MS	F	P
Sex	1	105.338	105.338	50.32	0.000
Maj	2	5.880	2.940	1.40	0.248
Sex*Maj	2	5.573	2.786	1.33	0.266
Error	228	477.282	2.093		
Total	233	594.073			

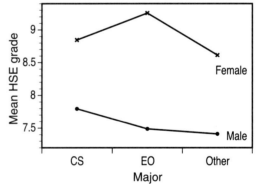

Gender		Major	
	CS	EO	Other
Male	$n = 39$	39	39
	$\bar{x} = 7.7949$	7.4872	7.4103
	$s = 1.5075$	2.1505	1.5681
Female	$n = 39$	39	39
	$\bar{x} = 8.8462$	9.2564	8.6154
	$s = 1.1364$	0.7511	1.1611

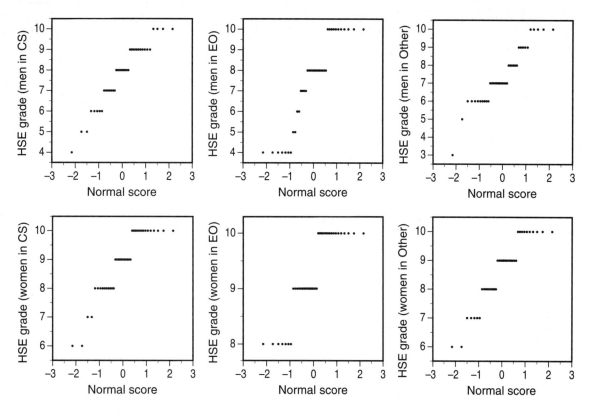

13.51. The table and plot of the means below suggest that students who stay in the sciences have higher mean SATV scores than those who end up in the "Other" group. Female CS and EO students have higher scores than males in those majors, but males have the higher mean in the Other group. Normal quantile plots suggest some right-skewness in the "Women in CS" group and also some non-Normality in the tails of the "Women in EO" group. Other groups look reasonably Normal. In the ANOVA table, only the effect of major is significant.

Minitab output: Two-way ANOVA for SATV on sex and major

Source	DF	SS	MS	F	P
Sex	1	3824	3824	0.47	0.492
Maj	2	150723	75362	9.32	0.000
Sex*Maj	2	29321	14661	1.81	0.166
Error	228	1843979	8088		
Total	233	2027848			

Gender		CS	EO	Other
Male	$n =$	39	39	39
	$\bar{x} =$	526.949	507.846	487.564
	$s =$	100.937	57.213	108.779
Female	$n =$	39	39	39
	$\bar{x} =$	543.385	538.205	465.026
	$s =$	77.654	102.209	82.184

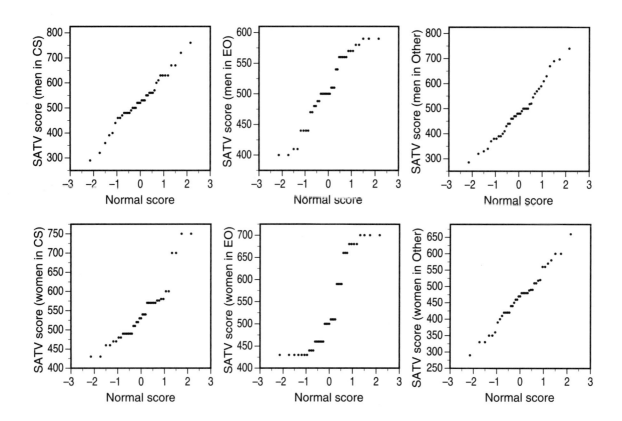

Chapter 14 Solutions

14.1. If $p = 0.5$, then odds $= \dfrac{p}{1 - p} = \dfrac{0.5}{0.5} = 1$, or "1 to 1."

14.3. We have $\hat{p}_{\text{men}} = \frac{63}{110} \doteq 0.5727$, and $\hat{p}_{\text{women}} = \frac{60}{130} = \frac{6}{13} \doteq 0.4615$. Therefore:

$$\text{odds}_{\text{men}} = \frac{63/110}{47/110} = \frac{63}{47} \doteq 1.3404, \text{ and}$$

$$\text{odds}_{\text{women}} = \frac{6/13}{7/13} = \frac{6}{7} \doteq 0.8571, \text{ or "6 to 7"}$$

> **Note:** *The odds can also be computed without first finding $\hat{p}$; for example, 63 men preferred Commercial A and 47 preferred Commercial B, so $\text{odds}_{\text{men}} = \frac{63}{47}$.*

14.5. With $\text{odds}_{\text{men}} = \frac{63}{47}$ and $\text{odds}_{\text{women}} = \frac{6}{7}$, we have $\log(\text{odds}_{\text{men}}) \doteq 0.2930$ and $\log(\text{odds}_{\text{women}}) \doteq -0.1542$.

> **Note:** *You may wish to remind students to use the natural logarithm, called "ln" by Excel and most calculators. A student who mistakenly uses the common (base-10) logarithm instead of the natural logarithm will get 0.1272 and −0.0669 as answers.*

14.7. The model is $y = \log(\text{odds}) = \beta_0 + \beta_1 x$. If $x = 1$ for men and 0 for women, we need:

$$\log\left(\frac{p_{\text{men}}}{1 - p_{\text{men}}}\right) = \beta_0 + \beta_1 \quad \text{and} \quad \log\left(\frac{p_{\text{women}}}{1 - p_{\text{women}}}\right) = \beta_0$$

We estimate $b_0 = \log(\text{odds}_{\text{women}}) \doteq -0.1542$ and $b_1 = \log(\text{odds}_{\text{men}}) - b_0 \doteq 0.4471$, so the regression equation is $\log(\text{odds}) = -0.1542 + 0.4471x$.

If $x = 0$ for men and 1 for women, we estimate $b_0 = \log(\text{odds}_{\text{men}}) \doteq 0.2930$ and $b_1 = \log(\text{odds}_{\text{women}}) - b_0 \doteq -0.4471$, so the regression equation is $\log(\text{odds}) = 0.2930 - 0.4471x$.

The estimated odds ratio is either:

$$e^{0.4471} \doteq \frac{\text{odds}_{\text{men}}}{\text{odds}_{\text{women}}} \doteq 1.5638 \quad \text{if } x = 1 \text{ for men, or}$$

$$e^{-0.4471} \doteq \frac{\text{odds}_{\text{women}}}{\text{odds}_{\text{men}}} \doteq 0.6395 \quad \text{if } x = 1 \text{ for women}$$

14.9. **(a)** The appropriate test would be a chi-square test with df $= 5$. **(b)** The logistic regression model has no error term. **(c)** H_0 should refer to β_1 (the population slope) rather than b_1 (the estimated slope). **(d)** The interpretation of coefficients is affected by correlations among explanatory variables.

14.11. In each case, we compute

$$\log(\text{odds}) = -3.1658 + 1.3083x$$

and odds $= e^{\log(\text{odds})}$.

	$x = $ LOpening	log(odds)	odds
(a)	3.219	1.0456	2.8452
(b)	3.807	1.8149	6.1405
(c)	4.174	2.2950	9.9249

14.13. (a) For each column, divide the "yes" entry by the total to find $\hat{p}$. **(b)** For each $\hat{p}$, compute odds $= \dfrac{\hat{p}}{1-\hat{p}}$. **(c)** Finally, take log(odds).

$$\hat{p}_{low} = \tfrac{88}{1169} = 0.0753 \qquad \text{odds}_{low} \doteq 0.0814 \qquad \log(\text{odds}_{low}) \doteq -2.5083$$

$$\hat{p}_{high} = \tfrac{112}{1246} = 0.0899 \qquad \text{odds}_{high} \doteq 0.0988 \qquad \log(\text{odds}_{high}) \doteq -2.3150$$

14.15. (a) $b_0 = \log(\text{odds}_{low}) \doteq -2.5083$ and $b_1 = \log(\text{odds}_{high}) - \log(\text{odds}_{low}) \doteq 0.1933$. **(b)** The fitted model is $\log(\text{odds}) = -2.5083 + 0.1933x$. **(c)** The odds ratio is $\text{odds}_{high}/\text{odds}_{low} = e^{b_1} \doteq 1.2132$ (or $\frac{0.0988}{0.0814} \doteq 1.2132$). **(d)** The relative risk from Example 9.7 was 1.19—very close to this odds ratio.

14.17. Shown below is Minitab output. **(a)** The slope is significantly different from 0 ($z = 2.37$, $P = 0.018$), but the constant is not ($z = 0.38$, $P = 0.706$). **(b)** With $b_1 = 1.0127$, $\text{SE}_{b_1} \doteq 0.4269$, and $z^* = 1.96$, the 95% confidence interval for β_1 is 0.176 to 1.849. **(c)** Exponentiating gives the interval $e^{0.176} \doteq 1.19$ to $e^{1.849} \doteq 6.36$.

Minitab output

Predictor	Coef	SE Coef	Z	P	Ratio	Lower	Upper
Constant	0.143101	0.378932	0.38	0.706			
Exclusive							
Yes	1.01267	0.426920	2.37	0.018	2.75	1.19	6.36

14.19. With $b_1 \doteq 3.1088$ and $\text{SE}_{b_1} \doteq 0.3879$, the 99% confidence interval is $b_1 \pm 2.576\text{SE}_{b_1} \doteq b_1 \pm 0.9992$, or 2.1096 to 4.1080.

14.21. (a) $z = \frac{3.1088}{0.3879} \doteq 8.01$. **(b)** $z^2 \doteq 64.23$, which agrees with the value of X^2 given by SPSS and SAS. **(c)** The sketches are below. For both the Normal and chi-square distributions, the test statistics are quite extreme, consistent with the reported P-value.

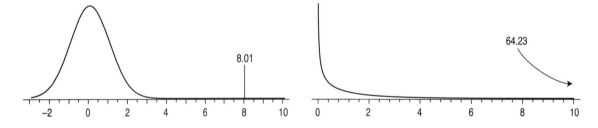

14.23. An odds ratio greater than 1 means a *higher* probability of a *low* tip. Therefore: The odds favor a low tip from senior adults, those dining on Sunday, those who speak English as a second language, and French-speaking Canadians. Diners who drink alcohol and lone males are less likely to leave low tips. For example, for a senior adult, the odds of leaving a low tip were 1.099 (for a probability of 0.5236).

14.25. (a) For men's magazines, the odds ratio confidence interval includes 1. This indicates that this explanatory variable has no effect on the probability that a model's clothing is not sexual, which is consistent with our failure to reject H_0 for men's magazines in the previous exercise. For all other explanatory variables, the odds ratio interval does not include 1, equivalent to the significant evidence against H_0 for those variables. **(b)** The odds that the model's clothing is not sexual are 1.27 to 2.16 times higher for magazines targeted at mature adults, 2.74 to 5.01 times higher when the model is male, and 1.11 to 2.23 times higher for magazines aimed at women. (These statements can also be made in terms of the odds that the model's clothing *is* sexual; for example, those odds are 1.27 to 2.16 times higher for magazines targeted at *young* adults, and so forth.) **(c)** The summary might note that it is easier to interpret the odds ratio rather than the regression coefficients because of the difficulty of thinking in terms of a log-odds scale.

14.27. (a) $\hat{p}_{hi} = \frac{73}{91} \doteq 0.8022$ and $odds_{hi} = \frac{\hat{p}_{hi}}{1-\hat{p}_{hi}} = 4.0\overline{5}$. **(b)** $\hat{p}_{non} = \frac{75}{109} \doteq 0.6881$ and $odds_{non} = \frac{\hat{p}_{non}}{1-\hat{p}_{non}} \doteq 2.2059$. **(c)** The odds ratio is $odds_{hi}/odds_{non} \doteq 1.8385$. The odds of a high-tech company offering stock options are about 1.84 times those for a non-high-tech firm.

14.29. (a) With $b_1 \doteq 0.6090$ and $SE_{b_1} \doteq 0.3347$, the 95% confidence interval is $b_1 \pm 1.96SE_{b_1} \doteq b_1 \pm 0.6560$, or -0.0470 to 1.2650. **(b)** Exponentiating the confidence limits gives the interval 0.9540 to 3.5430. **(c)** Because the confidence interval for β_1 contains 0, or equivalently because 1 is in the interval for the odds ratio, we could not reject H_0: $\beta_1 = 0$ at the 5% level. There does not appear to be a significant difference between the odds of stock options for high-tech and other firms.

Note: *Software reports $z = 1.820$ and a P-value of 0.0688, which are nearly identical to the results for a two-proportion z test with the same counts ($z = -1.832$ and $P = 0.0669$)—see the solution to Exercise 8.67. For large samples, these two tests should give similar results.*

Minitab output: Logistic regression (high-tech versus non-high-tech companies)

Predictor	Coef	SE Coef	Z	P	Odds Ratio	95% CI Lower	Upper
Constant	0.791128	0.206749	3.83	0.000			
HT							
Yes	0.608960	0.334663	1.82	0.069	1.84	0.95	3.54

14.31. (a) For the high blood pressure group, $\hat{p}_{hi} = \frac{55}{3338} \doteq 0.01648$, giving $odds_{hi} = \frac{\hat{p}_{hi}}{1-\hat{p}_{hi}} \doteq 0.01675$, or about 1 to 60. (If students give odds in the form "a to b," their choices of a and b might be different.) **(b)** For the low blood pressure group, $\hat{p}_{lo} = \frac{21}{2676} \doteq 0.00785$, giving $odds_{lo} = \frac{\hat{p}_{lo}}{1-\hat{p}_{lo}} \doteq 0.00791$, or about 1 to 126 (or 125). **(c)** The odds ratio is $odds_{hi}/odds_{lo} \doteq 2.1181$. Odds of death from cardiovascular disease are about 2.1 times greater in the high blood pressure group.

14.33. (a) The interval is $b_1 \pm 1.96SE_{b_1}$, or 0.2452 to 1.2558. **(b)** $X^2 = \left(\frac{0.7505}{0.2578}\right)^2 \doteq 8.47$. This gives a P-value between 0.0025 and 0.005. **(c)** We have strong evidence that there is a real (significant) difference in risk between the two groups.

14.35. (a) The estimated odds ratio is $e^{b_1} \doteq 2.1181$ (as we found in Exercise 14.31). Exponentiating the interval for β_1 in Exercise 14.33(a) gives the odds-ratio interval from about 1.28 to 3.51. **(b)** We are 95% confident that the odds of death from cardiovascular disease are about 1.3 to 3.5 times greater in the high blood pressure group.

Minitab output: Logistic regression on blood pressure

Predictor	Coef	SE Coef	Z	P	Odds Ratio	95% CI Lower	Upper
Constant	-4.83968	0.219078	-22.09	0.000			
BP							
hi	0.750498	0.257840	2.91	0.004	2.12	1.28	3.51

14.37. (a) The model is $\log\left(\frac{p_i}{1-p_i}\right) = \beta_0 + \beta_1 x_i$, where $x_i = 1$ if the ith person is over 40, and 0 if he/she is under 40. **(b)** p_i is the probability that the ith person is terminated; this model assumes that the probability of termination depends on age (over/under 40). In this case, that seems to have been the case, but we might expect that other factors were taken into consideration. **(c)** The estimated odds ratio is $e^{b_1} \doteq 3.859$. (Of course, we can also get this from $\frac{41/765}{7/504}$.) We can also find, for example, a 95% confidence interval for b_1: $b_1 \pm 1.96\mathrm{SE}_{b_1} = 0.5409$ to 2.1599. Exponentiating this translates to a 95% confidence interval for the odds: 1.7176 to 8.6701. The odds of being terminated are 1.7 to 8.7 times greater for those over 40. **(d)** Use a multiple logistic regression model, for example, $\log\left(\frac{p_i}{1-p_i}\right) = \beta_0 + \beta_1 x_{1,i} + \beta_2 x_{2,i}$.

14.39. It is difficult to find the needed probabilities from the numbers as given; this is made easier if we first convert the given information into a two-way table, shown on the right. The proportions meeting the activity guidelines are $\hat{p}_{\text{fruit}} = \frac{169}{237} \doteq 0.7131$ and $\hat{p}_{\text{no}} = \frac{494}{897} \doteq 0.5507$, so $\text{odds}_{\text{fruit}} \doteq$

Active?	Eats fruit? Yes	No	Total
Yes	169	494	663
No	68	403	471
Total	237	897	1134

2.4853 and $\text{odds}_{\text{no}} \doteq 1.2258$. Then $\log(\text{odds}_{\text{fruit}}) \doteq 0.9104$ and $\log(\text{odds}_{\text{no}}) \doteq 0.2036$, so $b_0 \doteq 0.2036$, $b_1 \doteq 0.7068$, and the model is $\log(\text{odds}) = 0.2036 + 0.7068x$. Software reports $\mathrm{SE}_{b_1} \doteq 0.1585$ and $z \doteq 4.46$ for testing H_0: $\beta_1 = 0$. A 95% confidence interval for the odds ratio is 1.49 to 2.77.

14.41. For each group, the probability, odds, and $\log(\text{odds})$ of being overweight are

$$\hat{p}_{\text{no}} = \frac{65080}{238215} \doteq 0.2732 \quad \text{odds}_{\text{no}} = \frac{\hat{p}_{\text{no}}}{1 - \hat{p}_{\text{no}}} \doteq 0.3759 \quad \log(\text{odds}_{\text{no}}) \doteq -0.9785$$

$$\hat{p}_{\text{FF}} = \frac{83143}{291152} \doteq 0.2856 \quad \text{odds}_{\text{FF}} = \frac{\hat{p}_{\text{FF}}}{1 - \hat{p}_{\text{FF}}} \doteq 0.3997 \quad \log(\text{odds}_{\text{FF}}) \doteq -0.9170$$

With $x = 0$ for no fast food and $x = 1$ for fast food, the logistic regression equation is $\log(\text{odds}) = -0.9785 + 0.0614x$. Software reports $\mathrm{SE}_{b_1} \doteq 0.006163$, and for testing H_0: $\beta_1 = 0$ we have $z \doteq 9.97$, leaving little doubt that the slope is not 0. A 95% confidence interval for the odds ratio is 1.0506 to 1.0763; the odds of being overweight for students at schools close to fast-food restaurants are about 1.05 to 1.08 times greater than for students at schools that are not close to fast food.

14.43. Portions of SAS and GLMStat output are given on the following page. **(a)** The X^2 statistic for testing this hypothesis is 33.65 (df = 3), which has $P = 0.0001$. We conclude that at least one coefficient is not 0. **(b)** The fitted model is $\log(\text{odds}) = -6.053 + 0.3710\,\text{HSM} + 0.2489\,\text{HSS} + 0.03605\,\text{HSE}$. The standard errors of the three coefficients are 0.1302, 0.1275, and 0.1253, giving respective 95% confidence intervals 0.1158 to 0.6262, -0.0010 to 0.4988, and -0.2095 to 0.2816. **(c)** Only the coefficient of

HSM is significantly different from 0, though HSS may also be useful.

Note: *In the multiple regression case study of Chapter 11, HSM was also the only significant explanatory variable among high school grades, and HSS was not even close to significant. See Figure 11.5 on page 603 of the text.*

SAS output

Criterion	Intercept Only	Intercept and Covariates	Chi-Square for Covariates
-2 LOG L	295.340	261.691	33.648 with 3 DF (p=0.0001)

Analysis of Maximum Likelihood Estimates

Variable	DF	Parameter Estimate	Standard Error	Wald Chi-Square	Pr > Chi-Square	Standardized Estimate
INTERCPT	1	-6.0528	1.1562	27.4050	0.0001	.
HSM	1	0.3710	0.1302	8.1155	0.0044	0.335169
HSS	1	0.2489	0.1275	3.8100	0.0509	0.233265
HSE	1	0.0361	0.1253	0.0828	0.7736	0.029971

GLMStat output

	estimate	se(est)	z ratio	Prob>|z|
1 Constant	-6.053	1.156	-5.236	<0.0001
2 HSM	0.3710	0.1302	2.849	0.0044
3 HSS	0.2489	0.1275	1.952	0.0509
4 HSE	3.605e-2	0.1253	0.2877	0.7736

14.45. The coefficients and standard errors for the fitted model are on the following page. Note that the tests requested in parts (a) and (b) are not available with all software packages. **(a)** The X^2 statistic for testing this hypothesis is given by SAS as 19.2256 (df = 3); because $P = 0.0002$, we reject H_0 and conclude that high school grades add a significant amount to the model with SAT scores. **(b)** The X^2 statistic for testing this hypothesis is 3.4635 (df = 2); because $P = 0.1770$, we cannot reject H_0; SAT scores do not add significantly to the model with high school grades. **(c)** For modeling the odds of HIGPA, high school grades (specifically HSM, and to a lesser extent HSS) are useful, while SAT scores are not.

SAS output

Analysis of Maximum Likelihood Estimates

Variable	DF	Parameter Estimate	Standard Error	Wald Chi-Square	Pr > Chi-Square	Standardized Estimate
INTERCPT	1	-7.3732	1.4768	24.9257	0.0001	.
HSM	1	0.3427	0.1419	5.8344	0.0157	0.309668
HSS	1	0.2249	0.1286	3.0548	0.0805	0.210704
HSE	1	0.0190	0.1289	0.0217	0.8829	0.015784
SATM	1	0.000717	0.00220	0.1059	0.7448	0.034134
SATV	1	0.00289	0.00191	2.2796	0.1311	0.147566

Linear Hypotheses Testing

Label	Wald Chi-Square	DF	Pr > Chi-Square
HS	19.2256	3	0.0002
SAT	3.4635	2	0.1770

GLMStat output

		estimate	se(est)	z ratio	Prob>|z|
1	Constant	-7.373	1.477	-4.994	<0.0001
2	SATM	7.166e-4	2.201e-3	0.3255	0.7448
3	SATV	2.890e-3	1.914e-3	1.510	0.1311
4	HSM	0.3427	0.1419	2.416	0.0157
5	HSS	0.2249	0.1286	1.748	0.0805
6	HSE	1.899e-2	0.1289	0.1473	0.8829

14.47. The models reported below are for the odds of *death*, as requested in the instructions. If a student models odds of survival, or codes the indicator variables for hospital and condition differently, his or her answers will be slightly different from these (but the conclusions should be the same). **(a)** The fitted model is $\log(\text{odds}) = -3.892 + 0.4157$ Hospital, using 1 for Hospital A and 0 for Hospital B. With $b_1 \doteq -0.4157$ and $\text{SE}_{b_1} \doteq 0.2831$, we find that $z = -1.47$ or $X^2 = 2.16$ ($P = 0.1420$), so we do not have evidence to suggest that β_1 is not 0. A 95% confidence interval for β_1 is -0.1392 to 0.9706 (this interval includes 0). We estimate the odds ratio to be $e^{b_1} \doteq 1.515$, with confidence interval 0.87 to 2.64 (this includes 1, since β_1 might be 0). **(b)** The fitted model is $\log(\text{odds}) = -3.109 - 0.1320$ Hospital $- 1.266$ Condition; as before, use 1 for Hospital A and 0 for Hospital B, 1 for good condition and 0 for poor. The estimated odds ratio is $e^{b_1} \doteq 0.8764$, with confidence interval 0.48 to 1.60. **(c)** In neither case is the effect of Hospital significant. However, we can see the effect of Simpson's paradox in the coefficient of Hospital, or equivalently in the odds ratio. In the model with Hospital alone, this coefficient was positive and the odds ratio was greater than 1, meaning Hospital A patients have higher odds of death. When condition is added to the model, this coefficient is negative and the odds ratio is less than 1, meaning Hospital A patients have lower odds of death.

GLMStat output: Hospital only

		estimate	se(est)	z ratio	Prob>|z|
1	Constant	-3.892	0.2525	-15.41	<0.0001
2	Hosp	0.4157	0.2831	-1.469	0.1420

		odds ratio	lower 95% ci	upper 95% ci
1	Constant	2.041e-2	1.244e-2	3.348e-2
2	Hosp	1.515	0.8701	2.639

Hospital and condition

		estimate	se(est)	z ratio	Prob>|z|
1	Constant	-3.109	0.2959	-10.51	<0.0001
2	Hosp	-0.1320	0.3078	-0.4288	0.6681
3	Cond	-1.266	0.3218	-3.935	<0.0001

		odds ratio	lower 95% ci	upper 95% ci
1	Constant	4.463e-2	2.499e-2	7.971e-2
2	Hosp	0.8764	0.4794	1.602
3	Cond	0.2820	0.1501	0.5298

Chapter 15 Solutions

15.1. The rankings are shown on the right. Group A ranks (shaded) are 1, 2, 4, 6, and 8; Group B ranks are 3, 5, 7, 9, and 10.

Group	Rooms	Ranks
A	30	1
A	68	2
B	240	3
A	243	4
B	329	5
A	448	6
B	540	7
A	552	8
B	560	9
B	780	10

15.3. The null hypothesis is H_0: no difference in distribution of number of rooms. The alternative might be two-sided ("there is a difference") or one-sided if we had a prior suspicion that one group had more rooms than the other. The test statistic is $W = 1 + 2 + 4 + 6 + 8 = 21$.

15.5. Under the null hypothesis,

$$\mu_W = \frac{(5)(11)}{2} = 27.5 \text{ and } \sigma_W = \sqrt{\frac{(5)(5)(11)}{12}} \doteq 4.7871$$

We found $W = 21$, so $z = \frac{21 - 27.5}{4.7871} \doteq -1.36$, for which the two-sided P-value is $2P(Z \le -1.36) \doteq 0.1738$. With the continuity correction, we find $z = \frac{21.5 - 27.5}{4.787} \doteq -1.25$, which gives $P = 2P(Z \le -1.25) = 0.2112$. The Minitab output below gives a similar P-value to that found with the continuity correction; the difference is due to the rounding of z. Regardless of the P-value used, we do not reject H_0.

   Note: *If a one-sided alternative was specified in Exercise 15.3, the P-value would be half as big: $P \doteq 0.0869$, or 0.1056 with the continuity correction.*

   In the Minitab output, the medians are referred to as ETA1 and ETA2 ("eta" is the Greek letter η). Minitab also reports an estimate of 0.21 for the difference $\eta_1 - \eta_2$; note that this is not the same as the difference between the two sample medians (0.7 − 0.4 = 0.3). This estimate, called the Hodges-Lehmann estimate, is not discussed in the text and has been removed from the Minitab outputs accompanying other solutions for this chapter. Briefly, this estimate is found by taking every response from the first group and subtracting every response from the second group, yielding (in this case) a total of 25 differences. The median of this set of differences is the Hodges-Lehmann estimate.

Minitab output: Wilcoxon rank sum (Mann-Whitney) confidence interval and test

```
GrpA       N =   5     Median =       243.0
GrpB       N =   5     Median =       540.0
Point estimate for ETA1-ETA2 is      -228.0
96.3 Percent C.I. for ETA1-ETA2 is (-537.0,208.0)
W = 21.0
Test of ETA1 = ETA2  vs.  ETA1 ~= ETA2 is significant at 0.2101
```

15.7. (a) For example, $\frac{99}{254} \doteq 39.0\%$ of self-employed workers are completely satisfied. The complete table is on the right with a bar graph. Overall, self-employed workers are more satisfied than the other group. **(b)** See the Minitab output below: $X^2 = 15.641$ with df $= 3$, for which $P = 0.001$. We can reject H_0 and conclude that job satisfaction and job type (self-employed or not) are not independent.

Self-employed?				
Yes	39.0%	55.9%	3.1%	2.0%
No	28.2%	61.2%	8.2%	2.3%

Minitab output: Chi-square test

```
Expected counts are printed below observed counts

          C2       C3       C4       C5     Total
    1     99      142        8        5       254
        77.83   152.53    18.06     5.58

    2    250      542       73       20       885
       271.17   531.47    62.94    19.42

Total  349      684       81       25      1139

ChiSq = 5.760 + 0.727 + 5.606 + 0.059 +
        1.653 + 0.209 + 1.609 + 0.017 = 15.641
df = 3, p = 0.001
```

15.9. Back-to-back stemplots on the right, summary statistics below. The men's distribution is skewed, and the women's distribution has a near-outlier. Men and women are not significantly different ($W = 1421$, $P = 0.6890$). The t test assumes Normal distributions; with small samples (like the previous exercise), this might be risky. In this exercise, the samples might be large enough to overcome the apparent non-Normality of the distributions.

Men		Women
42210	0	2
99999876655	0	5557777888889
221100	1	01223444
77665	1	66666777789
4331	2	0112244
965	2	
1	3	2
6	3	

Note: *Shown below is the Minitab output for a t test; the conclusion is the same as the Wilcoxon test ($t = -0.11$, $P = 0.92$).*

	$\bar{x}$	s	Min	Q_1	M	Q_3	Max
Men	14,060	9065	695	7464.5	11118	22740	36345
Women	14,252	6515	2363	8345.5	14602	18050	32291

Minitab output: Wilcoxon rank sum test

```
Mwords     N =  37      Median =        11118
Wwords     N =  41      Median =        14602
W = 1421.0
Test of ETA1 = ETA2  vs.  ETA1 ~= ETA2 is significant at 0.6890
```

Two-sample *t* test

```
            N      Mean     StDev    SE Mean
Mwords     37     14060      9065       1490
Wwords     41     14252      6515       1017

T-Test mu Mwords = mu Wwords (vs not =): T= -0.11  P=0.92  DF=  64
```

15.11. **(a)** Normal quantile plots are not shown. The score 0.00 for child 8 seems to be a low outlier (although with only five observations, such judgments are questionable). **(b)** For testing H_0: $\mu_1 = \mu_2$ versus H_a: $\mu_1 > \mu_2$, we have $\bar{x}_1 = 0.676$, $s_1 \doteq 0.1189$, $\bar{x}_2 = 0.406$, and $s_2 \doteq 0.2675$. Then, $t = 2.062$, which gives $P = 0.0447$ (df = 5.5). We have some evidence that high-progress readers have higher mean scores. **(c)** We test:

$$H_0: \text{Scores for both groups are identically distributed}$$
$$\text{vs. } H_a: \text{High-progress children systematically score higher}$$

for which we find $W = 36$ and $P \doteq 0.0473$ or 0.0463—significant evidence (at $\alpha = 0.05$) against the hypothesis of identical distributions. This is equivalent to the conclusion reached in part (b).

Minitab output: Wilcoxon rank sum test

```
HiProg1    N =   5      Median =        0.7000
LoProg1    N =   5      Median =        0.4000
W = 36.0
Test of ETA1 = ETA2  vs.  ETA1 > ETA2 is significant at 0.0473
The test is significant at 0.0463 (adjusted for ties)
```

15.13. **(a)** See table. **(b)** For Story 2, $W = 8 + 9 + 4 + 7 + 10 = 38$. Under H_0:

$$\mu_W = \frac{(5)(11)}{2} = 27.5$$

$$\sigma_W = \sqrt{\frac{(5)(5)(11)}{12}} \doteq 4.7871$$

(c) $z = \frac{38 - 27.5}{4.787} \doteq 2.19$; with the continuity correction, we compute $\frac{37.5 - 27.5}{4.787} \doteq 2.09$, which gives $P = P(Z > 2.09) = 0.0183$. **(d)** See the table.

Child	Progress	Story 1 Score	Story 1 Rank	Story 2 Score	Story 2 Rank
1	high	0.55	4.5	0.80	8
2	high	0.57	6	0.82	9
3	high	0.72	8.5	0.54	4
4	high	0.70	7	0.79	7
5	high	0.84	10	0.89	10
6	low	0.40	3	0.77	6
7	low	0.72	8.5	0.49	3
8	low	0.00	1	0.66	5
9	low	0.36	2	0.28	1
10	low	0.55	4.5	0.38	2

15.15. **(a)** At right. Unlogged plots appear to have a greater number of species. **(b)** We test H_0: There is no difference in the number of species on logged and unlogged plots versus H_a: Unlogged plots have a greater variety of species. The Wilcoxon test gives $W = 159$ and $P \doteq 0.0298$ (0.0290, adjusted for ties). We conclude that the observed difference is significant; unlogged plots really do have a greater number of species.

Unlogged		Logged
	0	4
	0	
	0	
	1	0
333	1	2
55	1	455
	1	7
998	1	88
10	2	
22	2	

Minitab output: Wilcoxon rank sum test

```
Unlogged   N =  12      Median =       18.500
Logged     N =   9      Median =       15.000
W = 159.0
Test of ETA1 = ETA2  vs.  ETA1 > ETA2 is significant at 0.0298
The test is significant at 0.0290 (adjusted for ties)
```

15.17. **(a)** We find $X^2 = 3.955$ with df $= (5 - 1)(2 - 1) = 4$, giving $P = 0.413$. There is little evidence to make us believe that there is a relationship between city and income. **(b)** Minitab reports $W = 56,370$, with $P \doteq 0.5$; again, there is no evidence that incomes are systematically higher in one city.

Minitab output: Wilcoxon rank sum test

```
City1    N = 241    Median =     2.0000
City2    N = 218    Median =     2.0000
W = 56370.0
Test of ETA1 = ETA2  vs.  ETA1 ~= ETA2 is significant at 0.5080
The test is significant at 0.4949 (adjusted for ties)
```

15.19. We test:

$$H_0: \text{Food scores and activities scores have the same distribution}$$
$$\text{vs. } H_a: \text{Food scores are higher}$$

The differences, and their ranks, are:

Spa	Food score	Activities score	Difference	Rank
1	77.3	95.7	−18.4	6
2	85.7	78.0	7.7	2
3	84.2	87.2	−3.0	1
4	85.3	85.3	0.0	–
5	83.7	93.6	−9.9	5
6	84.6	76.0	8.6	4
7	78.5	86.3	−7.8	3

The two positive differences have ranks 2 and 4, so $W^+ = 6$.

Note: *In assigning ranks, differences of 0 are ignored; see the comment in the text toward the bottom of page 735. If a student mistakenly assigns a rank of 1 to 0, they would find $W^+ = 3 + 5 = 8$ (or perhaps 9 if they erroneously count 0 as a "positive difference").*

15.21. Because one difference was 0, we ignore it and take $n = 6$, so that:

$$\mu_{W^+} = \frac{6(6+1)}{4} \doteq 10.5, \quad \sigma_{W^+} = \sqrt{\frac{6(6+1)(12+1)}{24}} = \sqrt{22.75} \doteq 4.7697,$$

and the approximate P-value is $P(W^+ \geq 5.5) \doteq P(Z \geq -1.05) \doteq 0.8531$. This agrees with the Minitab output below.

Note: *If a student does not see the instruction about discarding differences of 0 at the bottom of page 735, they might compute the mean and standard deviation using $n = 7$:*
$\mu_{W^+} = \frac{7(7+1)}{4} \doteq 14$ *and* $\sigma_{W^+} = \sqrt{\frac{7(7+1)(14+1)}{24}} = \sqrt{35} \doteq 5.9161$. *Such a student would presumably take $W^+ = 8$ (or 9), so they would compute the approximate P-value as* $P(W^+ \geq 7.5) \doteq P(Z \geq -1.10) \doteq 0.8643$ *or* $P(W^+ \geq 8.5) \doteq P(Z \geq -0.93) \doteq 0.8238$. *While these are close to the right answer (and lead to the same conclusion), they are not quite correct. In other situations, failing to ignore differences of 0 may lead to the wrong conclusion.*

Minitab output: Wilcoxon signed rank test (median = 0 versus median > 0)

	N	N FOR TEST	WILCOXON STATISTIC	P-VALUE	ESTIMATED MEDIAN
Diff	7	6	6.0	0.853	-3.450

15.23. (a) With this additional subject, six of the seven subjects rated drink A higher, and (as before) the subject who preferred drink B only gave it a 2-point edge. **(b)** For testing $H_0: \mu_d = 0$ versus $H_a: \mu_d \neq 0$, we have $\bar{x} \doteq 7.8571$ and $s \doteq 10.3187$, so $t \doteq 2.01$ (df = 6) and $P \doteq 0.0906$. **(c)** For testing H_0: Ratings have the same distribution for both drinks versus H_a: One drink is systematically rated higher, we have $W^+ = 26.5$ and $P = 0.043$. **(d)** The new data point is an outlier (see the stemplot, above on the right), which may make the t procedure inappropriate. This also increases the standard deviation of the differences, which makes t insignificant. The Wilcoxon test is not sensitive to outliers, and the extra data point makes it powerful enough to reject H_0.

```
-0 | 2
 0 | 2
 0 | 5578
 1 |
 1 |
 2 |
 2 |
 3 | 0
```

Minitab output: Matched pairs t test						
Variable	N	Mean	StDev	SE Mean	T	P-Value
Diff	7	7.86	10.32	3.90	2.01	0.091

Wilcoxon signed rank test				
		N FOR	WILCOXON		ESTIMATED
	N	TEST	STATISTIC	P-VALUE	MEDIAN
Diff	7	7	26.5	0.043	5.500

15.25. We examine the heart-rate increase (final minus resting) from low-rate exercise; our hypotheses are H_0: median $= 0$ versus H_a: median > 0. The statistic is $W^+ = 10$ (the first four differences are positive, and the fifth is 0, so we drop it). We compute
$$P = P(W^+ \geq 9.5) = P\left(\frac{W^+ - 5}{2.739} \geq \frac{9.5 - 5}{2.739}\right) \doteq P(Z \geq 1.64) = 0.0505.$$ This is right on the borderline of significance: It is fairly strong evidence that heart rate increases, but (barely) not significant at 5%.

Minitab output: Wilcoxon signed rank test (median = 0 versus median > 0)				
		N FOR	WILCOXON		ESTIMATED
	N	TEST	STATISTIC	P-VALUE	MEDIAN
LowDiff	5	4	10.0	0.050	7.500

15.27. For testing H_0: median $= 0$ versus H_a: median > 0, the Wilcoxon statistic is $W^+ = 119$ (14 of the 15 differences were positive, and the one negative difference was the smallest in absolute value), and $P < 0.0005$—very strong evidence that there are more aggressive incidents during moon days. This agrees with the results of the t and permutation tests.

Minitab output: Wilcoxon signed rank test (median = 0 versus median > 0)				
		N FOR	WILCOXON		ESTIMATED
	N	TEST	STATISTIC	P-VALUE	MEDIAN
diff	15	15	119.0	0.000	2.570

15.29. (a) At right. The distribution is clearly right-skewed but has no outliers. **(b)** $W^+ = 31$ (only 4 of 12 differences were positive) and $P = 0.556$—there is no evidence that the median is other than 105.

```
 9 | 1
 9 | 5679
10 | 134
10 | 5
11 | 1
11 | 9
12 | 2
```

Minitab output: Wilcoxon signed rank test (median = 105 versus median ≠ 105)				
		N FOR	WILCOXON		ESTIMATED
	N	TEST	STATISTIC	P-VALUE	MEDIAN
Radon	12	12	31.0	0.556	103.2

15.31. (a) The Wilcoxon statistic is $W^+ = 0$ (all of the differences were less than 16), for which $P = 0$—very strong evidence against H_0. We conclude that the median weight gain is less than 16 pounds. (b) Minitab (output below) gives the interval 3.75 to 5.90 kg for the median weight gain. (For comparison, in the solution to Exercise 7.32, the 95% confidence interval for the mean μ was about 3.80 to 5.66 kg.)

Minitab output: Wilcoxon signed rank test (median = 16 versus median $\neq$ 16)				
		N FOR	WILCOXON		ESTIMATED
	N	TEST	STATISTIC	P-VALUE	MEDIAN
Diff	16	16	0.0	0.000	4.800

Wilcoxon signed rank confidence interval				
		ESTIMATED	ACHIEVED	
	N	MEDIAN	CONFIDENCE	CONFIDENCE INTERVAL
Diff	16	4.80	94.8	(3.75, 5.90)

15.33. (a) For testing

H_0: The distribution of BMD is the same for all three groups

vs. H_a: At least one group is systematically higher or lower

we find $H = 9.10$ with df $= 2$, for which $P = 0.011$. (b) In the solution to Exercise 12.39, ANOVA yielded $F = 7.72$ (df 2 and 42) and $P = 0.001$. The ANOVA evidence is slightly stronger, but (at $\alpha = 0.05$) the conclusion is the same.

Minitab output: Kruskal-Wallis test				
LEVEL	NOBS	MEDIAN	AVE. RANK	Z VALUE
1	15	0.2190	20.1	-1.05
2	15	0.2160	17.7	-1.93
3	15	0.2320	31.2	2.97
OVERALL	45		23.0	

H = 9.10 d.f. = 2 p = 0.011
H = 9.12 d.f. = 2 p = 0.011 (adjusted for ties)

15.35. (a) Diagram on the following page. (b) The stemplots (right) suggest greater density for high-jump rats and a greater spread for the control group. (c) $H = 10.66$ with $P = 0.005$. We conclude that bone density differs among the groups. ANOVA tests H_0: all means are equal, assuming Normal distributions with the same standard deviation. For Kruskal-Wallis, the null hypothesis is that the distributions are the same (but not necessarily Normal). (d) There is strong evidence that the three groups have different bone densities; specifically, the high-jump group has the highest average rank (and the highest density), the low-jump group is in the middle, and the control group is lowest.

Control		Low jump		High jump	
55	4	55		55	
56	9	56		56	
57		57		57	
58		58	8	58	
59	33	59	469	59	
60	03	60	57	60	
61	14	61		61	
62	1	62		62	2266
63		63	1258	63	1
64		64		64	33
65	3	65		65	00
66		66		66	
67		67		67	4

Minitab output: Kruskal-Wallis test

LEVEL	NOBS	MEDIAN	AVE. RANK	Z VALUE
Ctrl	10	601.5	10.2	-2.33
Low	10	606.0	13.6	-0.81
High	10	637.0	22.6	3.15
OVERALL	30		15.5	

$H = 10.66$ d.f. $= 2$ $p = 0.005$
$H = 10.68$ d.f. $= 2$ $p = 0.005$ (adjusted for ties)

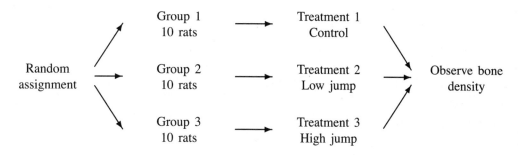

15.37. (a) $I = 4$, $n_i = 6$, $N = 24$. (b) The table below lists color, insect count, and rank. There are only three ties (and the second could be ignored, as both of those counts are for white boards). The R_i (rank sums) are:

$$
\begin{array}{llllllll}
\text{Yellow} & 17 & +20 & +21 & +22 & +23 & +24 & = 127 \\
\text{White} & 3 & +4 & +5.5 & +9.5 & +9.5 & +12.5 & = 44 \\
\text{Green} & 7 & +14 & +15 & +16 & +18 & +19 & = 89 \\
\text{Blue} & 1 & +2 & +5.5 & +8 & +11 & +12.5 & = 40
\end{array}
$$

(c) $H = \dfrac{12}{24(25)} \left(\dfrac{127^2 + 44^2 + 89^2 + 40^2}{6} \right) - 3(25) = 91.95\overline{3} - 75 = 16.95\overline{3}.$

Under H_0, this has approximately the chi-squared distribution with df $= I - 1 = 3$; comparing to this distribution tells us that $0.0005 < P < 0.001$.

B	B	W	W	W	B	G	B	W	W	B	W	B	G	G	G	Y	G	G	Y	Y	Y	Y	Y
7	11	12	13	14	14	15	16	17	17	20	21	21	25	32	37	38	39	41	45	46	47	48	59
1	2	3	4	5	6	7	8	9	10	11	12	13	14	15	16	17	18	19	20	21	22	23	24

5.5 (under 5, 6)   9.5 (under 9, 10)   12.5 (under 12, 13)

15.39. (a) Yes, the data support this statement: The percent of high-SES subject who have never smoked ($\frac{68}{211} \doteq 32.2\%$) is higher than those percents for middle- and low-SES subjects (17.3% and 23.7%, respectively), and the percent of current smokers among high-SES subjects ($\frac{51}{211} \doteq 24.2\%$) is lower than among the middle- (42.3%) and low- (46.2%) SES groups. **(b)** $X^2 = 18.510$ with df = 4, for which $P = 0.001$. There is a significant relationship between SES and smoking behavior. **(c)** $H = 12.72$ with df = 2, so $P = 0.002$—or, after adjusting for ties, $H = 14.43$ and $P = 0.001$. The observed differences are significant; some SES groups smoke systematically more.

Minitab output: Chi-square test

	Never	Former	Curr	Total
High	68	92	51	211
	58.68	83.57	68.75	
Mid	9	21	22	52
	14.46	20.60	16.94	
Low	22	28	43	93
	25.86	36.83	30.30	
Total	99	141	116	356

ChiSq = 1.481 + 0.850 + 4.584 +
         2.062 + 0.008 + 1.509 +
         0.577 + 2.119 + 5.320 = 18.510
df = 4, p = 0.001

Kruskal-Wallis test

LEVEL	NOBS	MEDIAN	AVE. RANK	Z VALUE
High	211	2.000	162.4	-3.56
Mid	52	2.000	203.6	1.90
Low	93	2.000	201.0	2.46
OVERALL	356		178.5	

H = 12.72   d.f. = 2   p = 0.002
H = 14.43   d.f. = 2   p = 0.001
                    (adjusted for ties)

15.41. (a) On the right is a histogram of service times for Verizon customers. With only 10 CLEC service calls, it is hardly necessary to make such a graph for them; we can simply observe that 7 of those 10 calls took 5 hours, which is quite different from the distribution for Verizon customers. The means and medians tell the same story:

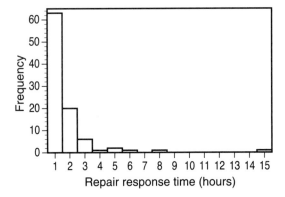

Verizon	$\bar{x}_V \doteq 1.7263$ hr	$M_V = 1$ hr
CLEC	$\bar{x}_C = 3.8$ hr	$M_C = 5$ hr

(b) The distributions are sharply skewed, and the sample sizes are quite different; the t test is not reliable in situations like this. The Wilcoxon rank-sum test gives $W = 4778.5$, which is highly significant ($P = 0.0026$ or 0.0006). We have strong evidence that response times for Verizon customers are shorter. It is also possible to apply the Kruskal-Wallis test (with two groups). While the P-values are slightly different ($P = 0.005$, or 0.001 adjusted for ties), the conclusion is the same: We have strong evidence of a difference in response times.

Minitab output: Wilcoxon rank sum test

```
Verizon     N = 95      Median =      1.000
CLEC        N = 10      Median =      5.000
W = 4778.5
Test of ETA1 = ETA2  vs.  ETA1 < ETA2 is significant at 0.0026
The test is significant at 0.0006 (adjusted for ties)
```

Kruskal-Wallis test

LEVEL	NOBS	MEDIAN	AVE. RANK	Z VALUE
1	95	1.000	50.3	-2.80
2	10	5.000	78.7	2.80
OVERALL	105		53.0	

```
H =  7.84  d.f. = 1  p = 0.005
H = 10.54  d.f. = 1  p = 0.001 (adjusted for ties)
```

15.43. See also the solutions to Exercises 1.79 and 12.35; the latter exercise requests the same analysis for ANOVA. The means, standard deviations, and medians (all in millimeters) are:

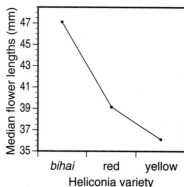

Variety	n	$\bar{x}$	s	M
bihai	16	47.5975	1.2129	47.12
red	23	39.7113	1.7988	39.16
yellow	15	36.1800	0.9753	36.11

The appropriate rank test is a Kruskal-Wallis test of H_0: all three varieties have the same length distribution versus H_a: at least one variety is systematically longer or shorter. We reject H_0 and conclude that at least one species has different lengths ($H = 45.35$, df $= 2$, $P < 0.0005$).

Minitab output: Kruskal-Wallis test

LEVEL	NOBS	MEDIAN	AVE. RANK	Z VALUE
1	16	47.12	46.5	5.76
2	23	39.16	26.7	-0.32
3	15	36.11	8.5	-5.51
OVERALL	54		27.5	

```
H = 45.35  d.f. = 2  p = 0.000
H = 45.36  d.f. = 2  p = 0.000 (adjusted for ties)
```

15.45. Use the Wilcoxon rank sum test with a two-sided alternative. For meat, $W = 15$ and $P = 0.4705$, and for legumes, $W = 10.5$ and $P = 0.0433$ (or 0.0421). There is no evidence of a difference in iron content for meat, but for legumes the evidence is significant at $\alpha = 0.05$.

Minitab output: Wilcoxon rank sum test for meat

```
Alum        N =   4     Median =       2.050
Clay        N =   4     Median =       2.375
W = 15.0
Test of ETA1 = ETA2  vs.  ETA1 ~= ETA2 is significant at 0.4705
Cannot reject at alpha = 0.05
```

Wilcoxon rank sum test for legumes

```
Alum        N =   4     Median =       2.3700
Clay        N =   4     Median =       2.4550
W = 10.5
Test of ETA1 = ETA2  vs.  ETA1 ~= ETA2 is significant at 0.0433
The test is significant at 0.0421 (adjusted for ties)
```

15.47. **(a)** The three pairwise comparisons are *bihai*-red, *bihai*-yellow, and red-yellow. **(b)** The test statistics and *P*-values are given in the Minitab output below; all *P*-values are reported as 0 to four decimal places. **(c)** All three are easily significant at the overall 0.05 level.

Minitab output: Wilcoxon rank sum test for *bihai* – red

```
bihai       N =  16     Median =      47.120
red         N =  23     Median =      39.160
W = 504.0
Test of ETA1 = ETA2  vs.  ETA1 ~= ETA2 is significant at 0.0000
```

Wilcoxon rank sum test for *bihai* – yellow

```
bihai       N =  16     Median =      47.120
yellow      N =  15     Median =      36.110
W = 376.0
Test of ETA1 = ETA2  vs.  ETA1 ~= ETA2 is significant at 0.0000
```

Wilcoxon rank sum test for red – yellow

```
red         N =  23     Median =      39.160
yellow      N =  15     Median =      36.110
W = 614.0
Test of ETA1 = ETA2  vs.  ETA1 ~= ETA2 is significant at 0.0000
```

Chapter 16 Solutions

The solutions for Chapter 16 present a special challenge. Because bootstrap and permutation methods require software, the answers will vary because of (a) random variation due to differences in resampling/rearrangement, and (b) possible systematic and feature differences arising from the specific software used.

Because of (a), most of the solutions here give *ranges* of possible answers, rather than a single answer. These ranges should include the results that most students should get from a single bootstrap or permutation run. (Basically, for each such exercise, I reported the minimum and maximum values from 1000 or more bootstraps or permutations.)

For (b), the text primarily refers to results from S-PLUS, but also mentions SAS and SPSS. If you have other statistical software, you can learn about its bootstrapping capabilities (if any) by consulting your document, or by doing a Web search for the name of your software and "bootstrap." Note that a free student version of S-PLUS is available at www.onthehub.com/tibco, so your students may use it for this chapter, even if they normally work with other software. (Faculty can download a 30-day evaluation copy.)

Many of these solutions were originally written by Tim Hesterberg (using S-PLUS) for earlier editions of *IPS,* and have been edited and updated by Darryl Nester using R, the free, open-source version of S-PLUS. One difference in R's bootstrapping library versus that of S-PLUS is that (at the time of this writing) it does not compute "tilting" confidence intervals, so those results are not given. If your software finds tilting intervals, they will (for most of these exercises) be similar to those found by other methods (percentile, BCa, etc.).

16.1. Student answers in this problem will vary substantially due to using different random numbers. (If they do not, you should be suspicious.) **(b)** While students could get a sample mean as low as 0, or as high as 29.78, 95% of all sample means should be between about 5 and 23. **(c)** Shown is a stemplot for a set of 200 resamples. Even for such a large number of resamples, the distribution is somewhat irregular; student stemplots (for 20 resamples) will be even more irregular. **(d)** The theoretical bootstrap standard error is about 4.694, but with only 20 resamples, there will be a fair amount of variation (although almost certainly in the range 2.9 to 6.5).

2	4
3	08
4	
5	09
6	2288
7	011133357
8	0022588
9	13477
10	0002355666889
11	1112233445557777788
12	00000012223333444566688
13	111111224555555688
14	0001144456667799
15	011225588888888889
16	0001444466677
17	133556668
18	02244478
19	0146799
20	001122447788
21	05
22	022225588
23	3
24	9
25	5

   Note: *The range of numbers (5 to 23) given in part (b) is based on 10000 resamples.*

   For part (d), the range of standard errors is based on the middle 99% of the SEs from 50000 separate resamples of size 20. The theoretical value is based on considering the six repair times as a population *to compute the standard deviation (dividing by $\sqrt{6}$ rather than $\sqrt{5}$), yielding $\sigma \doteq 11.497$, so the theoretical standard error is $\sigma/\sqrt{6} \doteq 4.694$. The computation in the text (page 16-6) does not mention this detail, although it is discussed briefly in Note 4 on page 16-57. Because bootstrap methods are generally not used with small samples, and the difference is negligible for large samples, it usually does not matter.*

16.3. (a) The bootstrap samples from the *sample* (that is, the *data*), not the *population*. **(b)** The bootstrap samples *with* replacement. **(c)** The sample size should be *equal to* the original sample. **(d)** The bootstrap distribution is usually similar to the sampling distribution *in shape and spread, but not in center.*

16.5. The bootstrap distribution is (usually) close to Normal, with some positive skewness. We expect the sampling distribution to be close to Normal.

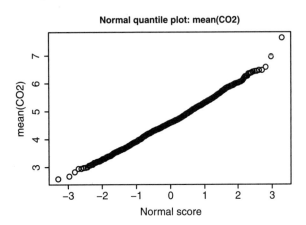
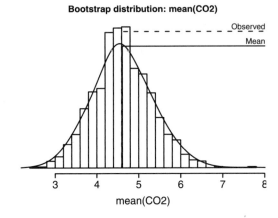

16.7. The bootstrap distribution suggests that the sampling distribution should be close to Normal.

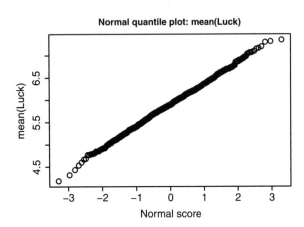
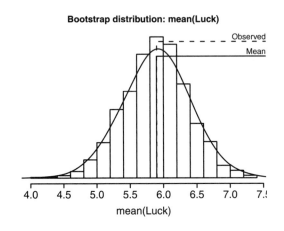

16.9. In each case, SE_{boot} will vary. To get an idea of how much variability one might observe, a range of "typical" bootstrap SE is given, based on 500 trials. **(a)** For the IQ data, $s \doteq 14.8009$, so $SE_{\bar{x}} \doteq 1.9108$. SE_{boot} will typically be between and 1.77 and 2.01. **(b)** For the CO_2 data, $s \doteq 4.8222$, so $SE_{\bar{x}} \doteq 0.6960$. SE_{boot} will typically be between about 0.64 and 0.74. **(c)** For the video-watching data, $s \doteq 3.8822$, so $SE_{\bar{x}} \doteq 1.3726$. SE_{boot} will typically be between about 1.20 and 1.36—almost certainly an underestimate. The bootstrap is biased downward for estimating standard error by a factor of about $\sqrt{(n-1)/n}$, which is about 0.94 when $n = 8$.

16.11. (a) The CALLCENTER20 bootstrap distribution is slightly skewed to the right, but it is considerably *less* skewed than the CALLCENTER80 bootstrap distribution. **(b)** The standard error for the smaller data set is much smaller: For CALLCENTER20, the standard error is almost always between 21 and 25, and for CALLCENTER80, it is almost always between 35 and 41.

Note: *The difference in standard errors is primarily because the sample standard deviation for the CALLCENTER20 data is much smaller (103.8 versus 342.0).*

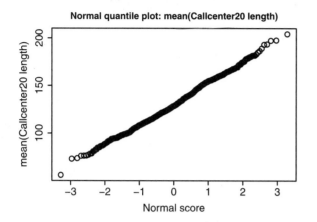
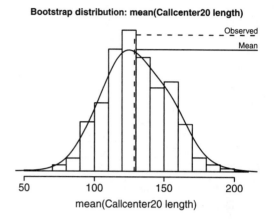

16.13. (a) The bootstrap distribution is skewed; a t interval might not be appropriate. (b) The bootstrap t interval is $\bar{x} \pm t^*SE_{boot}$, where $\bar{x} = 354.1$ sec, $t^* \doteq 2.0096$ for df = 49, and SE_{boot} is typically between 39.5 and 46.5. This gives the range of intervals shown on the right. (c) The interval reported in Example 7.11 was 266.6 to 441.6 seconds.

Typical ranges	
SE_{boot}	39.5 to 46.5
t lower	260.7 to 274.7
t upper	433.5 to 447.5

16.15. The summary statistics given in Example 16.6 include standard deviations $s_1 \doteq 14.7$ min for Verizon and $s_2 \doteq 19.5$ min for CLEC, so $SE_D \doteq 4.0820$. (Computation from the original data gives $SE_D \doteq 4.0827$.) The standard error reported by the S-PLUS bootstrap routine (shown in the text following that example) is 4.052.

16.17. (a) The bootstrap bias is typically between -4 and 4, which is small relative to $\bar{x} = 196.575$ min. (b) Ranges for the bootstrap interval are given on the right. (c) $SE_{\bar{x}} \doteq 38.2392$, while SE_{boot} ranges from about 35 to 41. The usual t interval is 120.46 to 272.69 min.

Typical ranges	
Bias	-4 to 4
SE_{boot}	35 to 41
t lower	114 to 127
t upper	266 to 279

16.19. The bootstrap distribution (following page) is noticeably non-Normal in the tails, especially the low tail. In addition, the bias is larger than we would like (and almost always negative, because of the heavy low tail). A t interval is risky (and perhaps not appropriate), but students may elect to compute it anyway.

Typical ranges	
Bias	-22.6 to -7.7
SE_{boot}	76.7 to 87.9
t lower	140.3 to 162.6
t upper	471.1 to 493.4

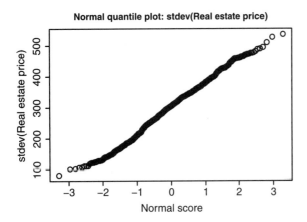

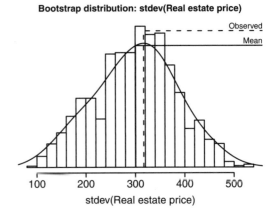

16.21. (a) The data appear to be roughly Normal, though with the typical random gaps and bunches that usually occur with relatively small samples. It appears from both the histogram and quantile plot that the mean is slightly larger than zero, but the difference is not large enough to rule out a $N(0, 1)$ distri-

Typical ranges	
Bias	−0.016 to 0.02
SE_{boot}	0.12 to 0.14
t lower	−0.16 to −0.11
t upper	0.36 to 0.41

bution. **(b)** The bootstrap distribution is extremely close to Normal with no appreciable bias. **(c)** $SE_{\bar{x}} \doteq 0.1308$, and the usual t interval is −0.1357 to 0.3854. Typical results for SE_{boot} and the bootstrap interval are above on the right.

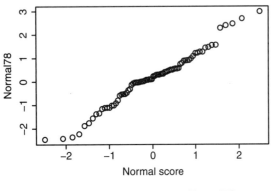

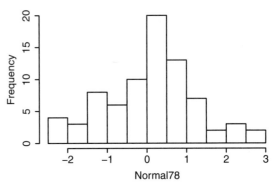

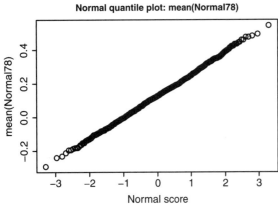

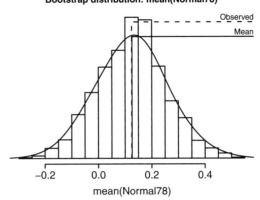

16.23. Because the scores are all between 1 and 10, there can be no extreme outliers, so standard *t* methods should be safe. The bootstrap distribution appears to be quite Normal, with little bias. The usual *t* interval is 4.9256 to 6.8744, which is in the range of typical bootstrap intervals.

Typical ranges	
Bias	−0.07 to 0.06
SE$_{boot}$	0.44 to 0.53
t lower	4.8 to 5.0
t upper	6.8 to 7.0

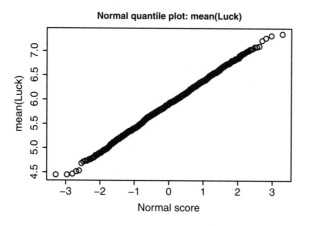

Normal quantile plot: mean(Luck)

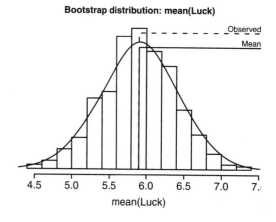

Bootstrap distribution: mean(Luck)

16.25. The distribution is sharply right-skewed, so a trimmed mean is a good choice. (Students will likely choose the 25% trimmed mean, both because it is mentioned in the text, and because it is given in the answer to this exercise.) For the sample, $\bar{x}_{25\%} = 1.74$ billion dollars. Typical ranges for the bias and SE$_{boot}$ are given on the right; note that the bias is a substantial fraction of SE$_{boot}$.

Typical ranges	
Bias	0.048 to 0.114
SE$_{boot}$	0.29 to 0.41
t lower	0.89 to 1.12
t upper	2.36 to 2.59

The bootstrap distribution is strongly right-skewed, so a bootstrap *t* interval would be questionable; the typical range of intervals is shown (above, right).

Note: *By definition, the mean wealth (trimmed or not) of billionaires must be more than $1 billion; the fact that the bootstrap interval can extend below that limit is an indication that we should not rely on it.*

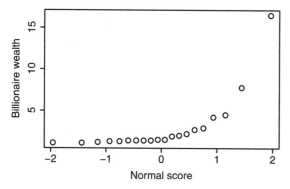

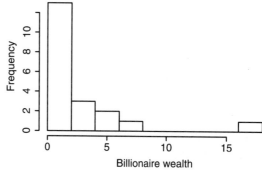

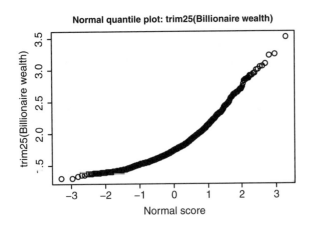
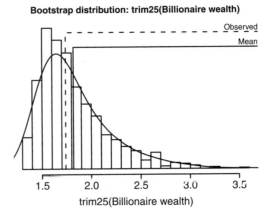

16.27. (a) $\bar{x}$ would have a Normal distribution with mean 8.4 and standard deviation $14.7/\sqrt{n}$. (b) and (c) Histograms below. The values of SE_{boot} will be quite variable, both because of variation in the original sample, and variation due to resampling. (d) Student answers will vary, depending on their samples. There may be some skewness (right or left) for smaller samples. SE_{boot} should be roughly halved each time the sample size increases by a factor of 4, although for $n = 10$ and $n = 40$, the size of SE_{boot} can vary considerably.

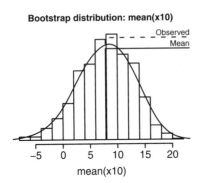
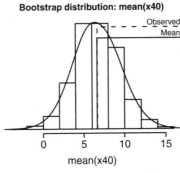
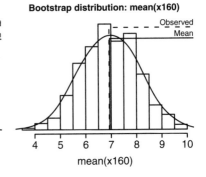

16.29. Student answers should vary depending on their samples. They should notice that the bootstrap distributions are approximately Normal for larger sample sizes. For small samples, the sample could be skewed one way or the other in Exercise 16.27, and most should be right skewed for Exercise 16.28. Some of that skewness should come through into the bootstrap distribution.

16.31. (a) The bootstrap distribution looks close to Normal (though that does not mean much with this small sample). The bias is small. (b) The typical range of bootstrap t intervals is on the right. (c) The bootstrap percentile interval is much narrower than the bootstrap t interval. (It is typically 70% to 80% as wide.)

Typical ranges	
Bias	−0.54 to 0.54
SE_{boot}	4.35 to 5.01
t lower	0.89 to 2.57
t upper	24.94 to 26.63
Percentile lower	3.7 to 5.9
Percentile upper	22.1 to 24.5

Note: *The reason that the percentile interval is narrower in this setting is that the bootstrap distribution has "heavy tails," visible in the slight curvature on the edges of the quantile plot. This inflates the standard error, and therefore make the t interval wider than it should be.*

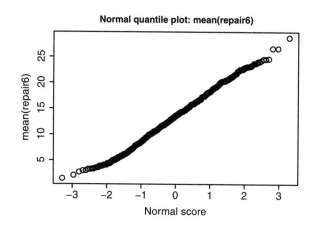

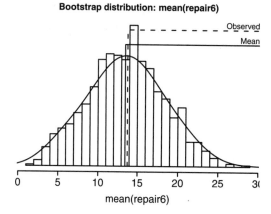

16.33. (a) The bootstrap percentile and *t* intervals are very similar, suggesting that the *t* intervals are acceptable. **(b)** Every interval (percentile and *t*) includes 0.

   Note: *In the solution to Exercise 16.31, the percentile intervals were always 70% to 80% as wide as the t intervals (because of the heavy tails of that bootstrap distribution). In this case, the width of the percentile interval is 93% to 106% of the width of the t interval.*

Typical ranges	
t lower	−0.16 to −0.11
t upper	0.36 to 0.41
Percentile lower	−0.17 to −0.09
Percentile upper	0.35 to 0.42

16.35. These intervals are given on page 16-37 of the text: The percentile interval is (−0.128, 0.356), and the bootstrap *t* interval is (−0.144, 0.358). The differences are relatively small relative to the width of the intervals, so they do not indicate appreciable skewness.

16.37. Typical ranges for the BCa interval are shown on the right; the tilting interval will be similar. Most intervals are fairly similar to the bootstrap *t* and percentile intervals from Example 16.10, suggesting that the simpler intervals are adequate.

Typical ranges	
BCa lower	−0.19 to −0.11
BCa upper	0.32 to 0.41

16.39. The percentile interval is shifted to the right relative to the bootstrap *t* interval. The more accurate intervals are shifted even further to the right.

Typical ranges	
t lower	114 to 127
t upper	266 to 279
Percentile lower	127 to 140
Percentile upper	267 to 298
BCa lower	137 to 152
BCa upper	292 to 371

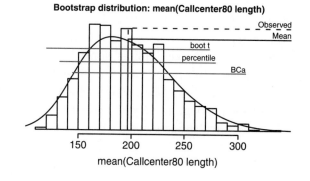

16.41. The bootstrap distribution for the smaller sample is *less* skewed. The standard *t* interval is 80.34 to 177.46; the bootstrap *t* interval is similar, and the other bootstrap intervals are generally narrower and shifted to the right.

Note: *Generally, a smaller sample should result in* less *regularity—that is, more skewness, larger standard error, etc. In this case, the smaller sample does not contain the nine highest call lengths, many of which would be considered outliers. Those increase the skewness of the bootstrap distribution for CALLCENTER80.*

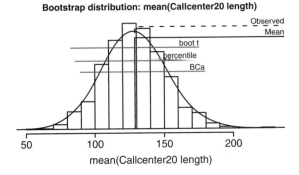

Typical ranges	
t lower	78.0 to 84.4
t upper	173.3 to 179.8
Percentile lower	80.8 to 92.8
Percentile upper	169.7 to 181.6
BCa lower	83.1 to 96.2
BCa upper	173.0 to 191.9

16.43. The observed difference is $\bar{x}_{ILEC} - \bar{x}_{CLEC} \doteq -8.1$. Ranges for all three intervals are given below. Because of the left skew of the bootstrap distribution, the *t* interval does not reach far enough to the left and reaches too far to the right, meaning that the interval would be too high too often, effectively overestimating where the true difference lies. This may also be true for the percentile interval, but considerably less so.

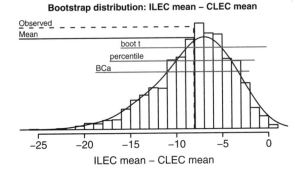

Typical ranges	
t lower	−16.5 to −15.4
t upper	−0.8 to 0.3
Percentile lower	−18.5 to −16.0
Percentile upper	−2.1 to −1.2
BCa lower	−19.3 to −16.5
BCa upper	−2.5 to −1.4

16.45. (a) The bootstrap distribution is sharply left-skewed. **(b)** Shown are ranges for the percentile and BCa intervals, as well as the (inappropriate) bootstrap *t* interval. The percentile and BCa intervals typically have similar upper limits, but the BCa lower limit is generally less than the percentile lower limit. The confidence intervals give more than enough evidence to reject H_0: $\rho = 0$; in fact, we have strong evidence that the correlation is at least 0.99.

Typical ranges	
Bias	−0.00019 to 0.00008
t lower	0.9940 to 0.9948
t upper	0.9995 to 1.0002
Percentile lower	0.9931 to 0.9946
Percentile upper	0.9988 to 0.9992
BCa lower	0.9899 to 0.9937
BCa upper	0.9985 to 0.9990

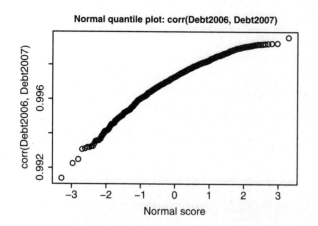

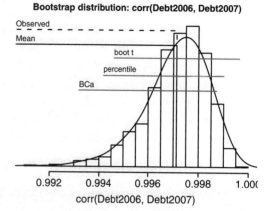

16.47. (a) The regression equation for predicting salary (in $millions) is $\hat{y} = 0.8125 + 7.7170x$, where x is batting average. (The slope is *not* significantly different from 0: $t = 0.744$, $P = 0.461$.) **(b)** The bootstrap distribution is reasonably close to Normal, suggesting that any of the intervals would be reasonably accurate. Ranges for the intervals are given on the right. **(c)** These results are consistent with our conclusion about correlation: All intervals include zero, which corresponds to no (linear) relationship between batting average and salary.

Typical ranges	
Bias	−0.56 to 1.35
SE_{boot}	9.26 to 10.6
t lower	−13.6 to −10.8
t upper	26.3 to 29.0
Percentile lower	−12.7 to −7.7
Percentile upper	26.0 to 31.7
BCa lower	−13.6 to −7.6
BCa upper	24.7 to 33.0

R output: `lm(I(Salary/10^6) ~ Average, data=mlbsalaries)`

```
Coefficients:
              Estimate Std. Error t value Pr(>|t|)
(Intercept)     0.8125     2.6941   0.302    0.764
Average         7.7170    10.3738   0.744    0.461

Residual standard error: 2.718 on 48 degrees of freedom
Multiple R-squared: 0.0114,     Adjusted R-squared: -0.009199
F-statistic: 0.5534 on 1 and 48 DF,  p-value: 0.4606
```

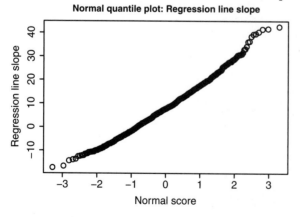

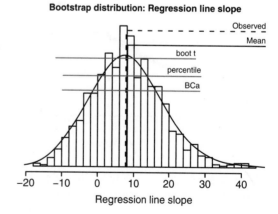

16.49. (a) The tails of the residuals are somewhat heavier than we would expect from a Normal distribution. (b) The bootstrap distribution is reasonable close to Normal, so the bootstrap t should be fairly accurate. (c) Ranges for all three bootstrap intervals are given in the table on the right; they all give similar results. With $t^* = 2.074$ for df $= 22$, the standard t interval is $1.1159 \pm t^*(0.0108) = 1.0784$ to 1.1534. The bootstrap intervals are fairly close to this.

Typical ranges	
Bias	-0.0008 to 0.0026
SE_{boot}	0.016 to 0.019
t lower	1.07 to 1.09
t upper	1.14 to 1.16
Percentile lower	1.08 to 1.09
Percentile upper	1.14 to 1.16
BCa lower	1.07 to 1.09
BCa upper	1.14 to 1.16

R output: `lm(formula = Debt2007 ~ Debt2006, data = debt)`

```
Coefficients:
             Estimate Std. Error t value Pr(>|t|)
(Intercept)  1.11113    3.60162    0.309    0.76
Debt2006     1.11594    0.01808   61.727   <2e-16

Residual standard error: 11.09 on 22 degrees of freedom
Multiple R-squared: 0.9943,     Adjusted R-squared: 0.994
F-statistic:  3810 on 1 and 22 DF,  p-value: < 2.2e-16
```

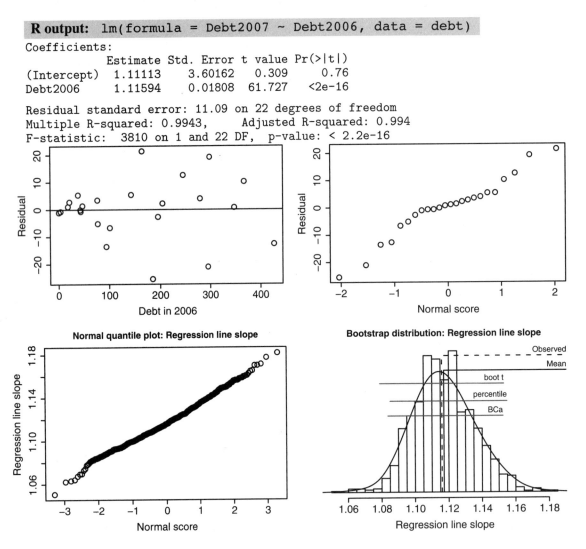

16.51. No, because we believe that one population has a smaller spread, but in order to pool the data, the permutation test requires that both populations be the same when H_0 is true.

16.53. (a) The observed difference in means is $\frac{57+53}{2} - 19 + 37 + 41 + 424 = 20.25$.

(b) Student results will vary, but should be one of the 15 (equally likely) possible values:

$$-20.25, \ -17.25, \ -16.5, \ -8.25, \ -5.25, \ -3.75, \ -3, \ 0, \ 5.25, \ 8.25, \ 9, \ 11.25, \ 12, \ 20.25$$

(c) The histogram shape will vary considerably. **(d)** Out of 20 resamples, the number which yield a difference of 20.25 or more has a binomial distribution with $n = 20$ and $p = 1/15$, so most students should get between 0 and 4, for a *P*-value between 0 and 0.2. **(e)** As was noted in part (b), only one resample gives a difference of means greater than or equal to the observed value, so the exact *P*-value is $1/15 \doteq 0.0667$.

> **Note:** *To determine the 15 possible values, note that the six numbers sum to 249. If the first two numbers add up to T, then the other four will add up to $249 - T$, and the difference in means will be $\frac{1}{2}T - \frac{1}{4}(249 - T) = \frac{3}{4}T - 62.25$. The values of T range from $19 + 37 = 56$ to $57 + 53 = 110$.*

16.55. (a) The ILEC distribution (gray bars) is clearly skewed to the right, while the CLEC distribution is skewed to the left (although with only 10 observations, that might not mean anything). **(b)** In keeping with the discussion in Example 16.13, we use a one-sided alternative. For a test of $H_0: \mu_{\text{ILEC}} = \mu_{\text{CLEC}}$ versus $H_a: \mu_{\text{ILEC}} < \mu_{\text{CLEC}}$, we find $t \doteq -3.25$, df $\doteq 10.71$, and $P \doteq 0.004$. **(c)** Based on

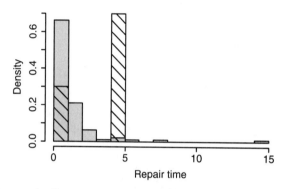

1000 resamples (each of size 1000), the *P*-value typically ranges from 0.001 to 0.010. The permutation test does not require Normal distributions, and gives more accurate answers in the case of skewness. A plot of the permutation distribution shows there is substantial skewness. **(d)** The difference is significant at the 5% level (and usually at the 1% level).

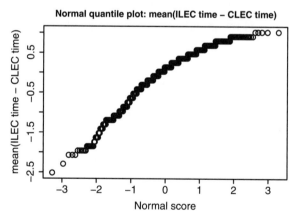

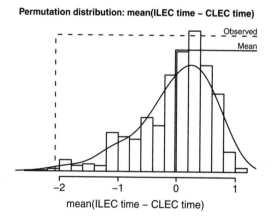

16.57. (a) The two populations should be the same shape, but skewed—or otherwise clearly non-Normal—so that the *t* test is not appropriate. **(b)** Either test is appropriate if the two populations are both Normal with the same standard deviation. **(c)** We can use a *t* test, but not a permutation test, if both populations are Normal with different standard deviations.

16.59. (a) We test H_0: $\mu = 0$ versus H_a: $\mu > 0$, where μ is the population mean difference before and after the summer language institute. We find $t \doteq 3.86$, df $= 19$, and $P \doteq 0.0005$. **(b)** The quantile plot (right) looks odd because we have a small sample, and all differences are integers. **(c)** The P-value is almost always less than 0.002. Both tests lead to the same conclusion: The difference is statistically significant.

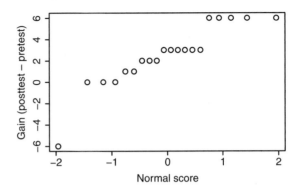

Note: *The text states that "the histogram [of the permutation distribution] looks a bit odd." In fact, different software produces different default histograms, some of which look fine. (This statement was made about the default histogram produced by S-PLUS.) To avoid potential confusion, check what your software does, and (if necessary) tell students to ignore that part of the question.*

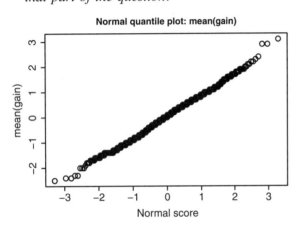

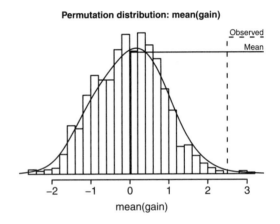

16.61. We test H_0: $\rho = 0$ versus H_a: $\rho \neq 0$, where ρ is the population correlation. (One could also justify the one-sided alternative $\rho > 0$ in this case; the ultimate conclusion is the same for either alternative.) The permutation distribution (found by permuting the debts from one year, then computing the correlation) is roughly Normal. In the histogram of the permutation distribution below, the observed correlation ($r \doteq 0.997$) is not marked because it lies far out on the high tail, nearly five standard deviations above the mean (0). Consequently, the reported P-value is nearly always 0, confirming the very strong evidence found in the solution to Exercise 16.45.

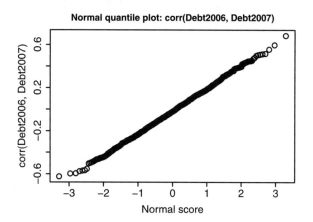

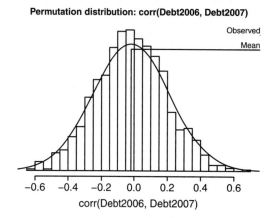

16.63. For testing H_0: $\sigma_1 = \sigma_2$ versus H_a: $\sigma_1 \ne \sigma_2$, the permutation test P-value will almost always be between 0.065 and 0.095. In the solution to Exercise 7.105, we found $F \doteq 1.50$ with df $= 29$ and 29, for which $P \doteq 0.2757$—three or four times as large. In this case, the permutation test P-value is smaller, which is typical of short-tailed distributions.

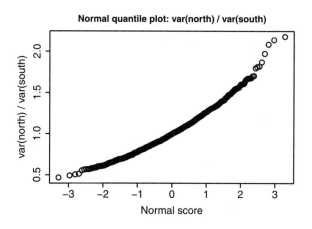

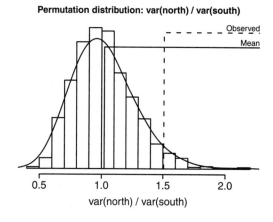

16.65. For the permutation test, we must resample in a way that is consistent with the null hypothesis. Hence we pool the data—assuming that the two populations are the same—and draw samples (without replacement) for each group from the pooled data. For the bootstrap, we do not assume that the two populations are the same, so we sample (with replacement) from each of the two datasets separately, rather than pooling the data first.

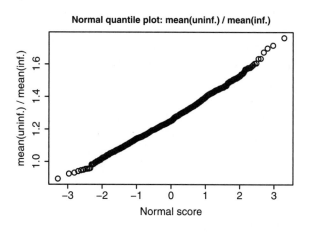

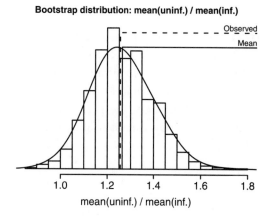

16.67. (a) The resampled CLEC standard deviation is sometimes 0, so for best results (that is, to avoid infinite ratios), put that standard deviation in the numerator. Both bootstrap distributions are shown below. (We do not need quantile plots to confirm that these distributions are non-Normal.) Regardless of which ratio we use, the resulting P-value is close to 0.37. **(b)** The difference in the P-values is evidence of the inaccuracy of the F test; these distributions clearly do not satisfy the Normality assumption.

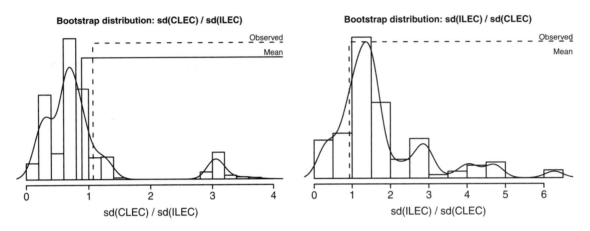

16.69. The bootstrap distribution looks quite Normal, and (as a consequence) all of the bootstrap confidence intervals are similar to each other, and also are similar to the standard (large-sample) confidence interval: 0.0981 to 0.1415.

   Note: *At the time these solutions were written, R's bootstrapping package would fail if asked to find the BCa confidence interval for this exercise.*

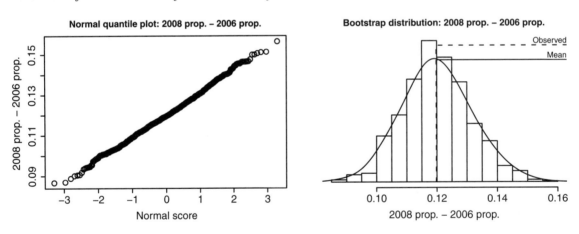

16.71. (a) The standard test of H_0: $\sigma_1 = \sigma_2$ versus H_a: $\sigma_1 \neq \sigma_2$ leads to $F = 0.3443$ with df 13 and 16; $P \doteq 0.0587$. **(b)** The permutation P-value is typically between 0.02 and 0.03. **(c)** The P-values are similar, even though technically, the permutation test is significant at the 5% level, while the standard test is (barely) not. Because the samples are too small to assess Normality, the permutation test is safer. (In fact, the population distributions are discrete, so they cannot follow Normal distributions.)

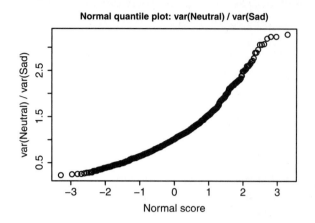

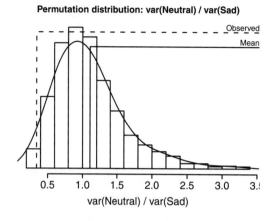

16.73. See the solution to Exercise 16.18 for another view of the bootstrap distribution. Ranges for the bootstrap t, percentile, and BCa intervals are given on the right. All have similar upper endpoints, but the lower endpoint for the bootstrap t is typically less than the others (because it ignores the skewness in the bootstrap distribution). It appears that any of the other intervals—including the percentile interval—would be more reliable.

Typical ranges	
t lower	205.5 to 211.5
t upper	276.5 to 282.5
Percentile lower	208.4 to 216.5
Percentile upper	276.2 to 288.8
BCa lower	207.9 to 217.4
BCa upper	276.0 to 293.6

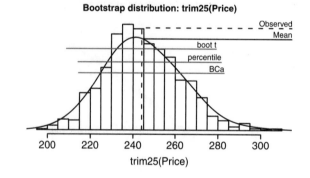

16.75. All answers (including the shape of the bootstrap distribution) will depend strongly on the initial sample of uniform random numbers. The median M of these initial samples will be between about 0.36 and 0.64 about 95% of the time; this is the center of the bootstrap t confidence interval. **(a)** For a uniform distribution on 0 to 1, the population median is 0.5. Most of the time, the bootstrap distribution is quite non-Normal; three examples are shown below. **(b)** SE_{boot} typically ranges from about 0.04 to 0.12 (but may vary more than that, depending on the original sample). The bootstrap t interval is therefore roughly $M \pm 2SE_{boot}$. **(c)** The more sophisticated BCa and tilting intervals may or may not be similar to the bootstrap t interval. The t interval is not appropriate because of the non-Normal shape of the bootstrap distribution, and because SE_{boot} is unreliable for the sample median (it depends strongly on the sizes of the gaps between the observations near the middle).

Note: *Based on 5000 simulations of this exercise, the bootstrap t interval $M \pm 2SE_{boot}$ will capture the true median (0.5) only about 94% of the time (so it slightly underperforms its intended 95% confidence level). In the same test, both the percentile and BCa intervals included 0.5 over 95% of the time, while at the same time being narrower than the bootstrap t interval nearly two-thirds of the time. These two measures (achieved confidence level, and width of confidence interval) both confirm the superiority of the other intervals.*

The bootstrap percentile, BCa, and tilting intervals do fairly well despite the high variability in the shape of the bootstrap distribution. They give answers similar to the exact rank-based confidence intervals obtained by inverting hypothesis tests. One variation of tilting intervals matches the exact intervals.

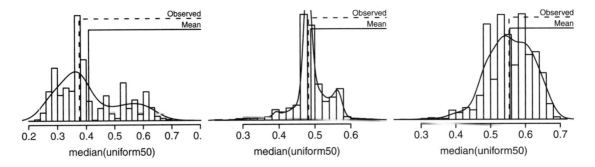

16.77. See the solution to Exercise 2.33 for a scatterplot. The permutation distribution (found by permuting one variable and computing the correlation) is roughly Normal, and the observed correlation ($r \doteq 0.878$) lies far out on the high tail, about three standard deviations above the mean (0). We conclude there is a significant positive relationship.

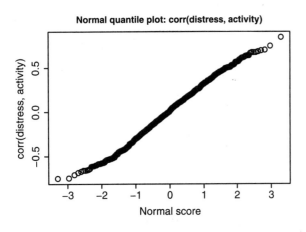

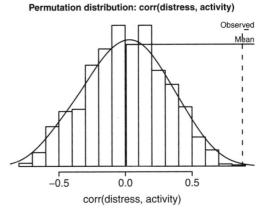

16.79. (a) The 2001 data is slightly skewed, but close to Normal given the sample size (50). The 2000 data is strongly right-skewed with two high outliers; a sample of size 20 is probably not enough to compensate. **(b)** The two-sided P-value for the permutation test is approximately 0.28. We conclude that there is not strong evidence that the mean selling prices are different for all Seattle real estate in 2000 and in 2001.

2000 prices		2001 prices	
1	3346899	0	5677
2	001488	1	0134445799
3	3669	2	0011123444677899
4	8	3	1123457
5		4	25556788
6		5	017
7		6	8
8		7	1
9			
10			
11	0		
12			
13			
14			
15			
16			
17			
18	4		

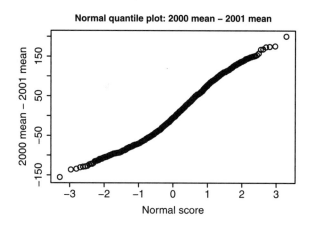

Normal quantile plot: 2000 mean – 2001 mean

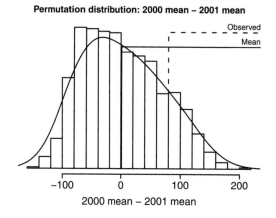

Permutation distribution: 2000 mean – 2001 mean

16.81. The permutation and bootstrap distributions for the difference in medians are extremely non-Normal, with many gaps and multiple peaks. In this situation, we have conflicting results: The permutation test gives fairly strong evidence of a difference (the two-sided P-value is roughly 0.032), but the BCa interval for the difference in medians nearly always includes 0.

Typical ranges	
t lower	9.6 to 16.5
t upper	93.5 to 100.4
Percentile lower	0 to 30
Percentile upper	90 to 100
BCa lower	−32.5 to 0
BCa upper	75 to 90

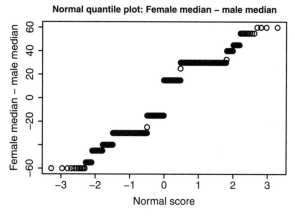

Normal quantile plot: Female median – male median

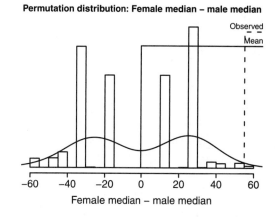

Permutation distribution: Female median – male median

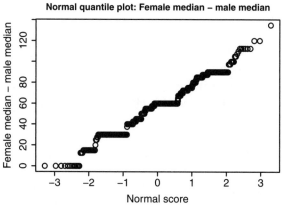

Normal quantile plot: Female median – male median

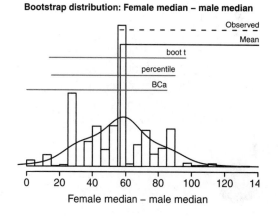

Bootstrap distribution: Female median – male median

16.83. See Exercise 8.55 for more details about this survey. The bootstrap distribution appears to be close to Normal; bootstrap intervals are similar to the large-sample interval (0.3146 to 0.3854).

Typical ranges	
t lower	0.31 to 0.32
t upper	0.38 to 0.39
Percentile lower	0.30 to 0.32
Percentile upper	0.38 to 0.39

Note: *At the time of this writing, R's bootstrapping package would not compute the BCa intervals for this exercise.*

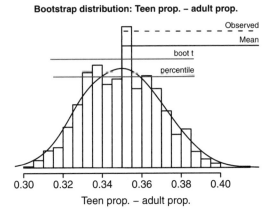

16.85. (a) This is the usual way of computing percent change: $89/54 - 1 = 0.65$. **(b)** Subtract 1 from the confidence interval found in Exercise 16.84; this typically gives an interval similar to 0.55 to 0.75. **(c)** Preferences will vary.

16.87. (a) The mean ratio is 1.0596; the usual *t* interval is $1.0596 \pm (2.262)(0.02355) \doteq 1.0063$ to 1.1128. The bootstrap distribution for the mean is close to Normal, and the bootstrap confidence intervals (typical ranges on the right) are usually similar to the usual *t* interval, but slightly narrower. Bootstrapping the median produces a clearly non-Normal distribution; the bootstrap *t* interval should not be used for the median. (Ranges for median intervals are not given.) **(b)** The ratio of means is 1.0656; the bootstrap distribution is noticeably skewed, so the bootstrap *t* is not a good choice, but the other methods usually give intervals similar to 0.75 to 1.55. Also shown below is the bootstrap distribution for the ratio of the medians. It is considerably less erratic than the median ratio,

Typical ranges	
(a) Mean ratio	
t lower	1.00 to 1.02
t upper	1.10 to 1.12
Percentile lower	1.00 to 1.03
Percentile upper	1.09 to 1.11
BCa lower	1.00 to 1.03
BCa upper	1.09 to 1.11
(b) Ratio of means	
t lower	0.59 to 0.68
t upper	1.46 to 1.54
Percentile lower	0.69 to 0.78
Percentile upper	1.45 to 1.64
BCa lower	0.69 to 0.78
BCa upper	1.45 to 1.66

but we have still not included these confidence intervals. **(c)** For example, the usual *t* interval from part (a) could be summarized by the statement, "On average, Jocko's estimates are 1% to 11% higher than those from other garages."

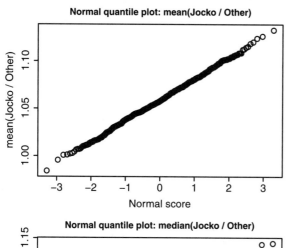

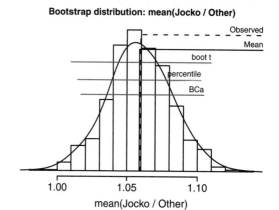

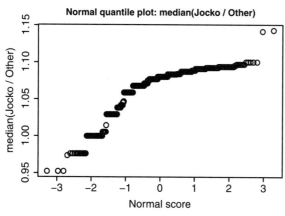

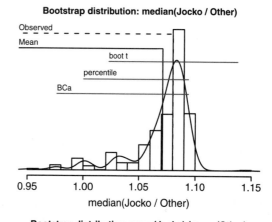

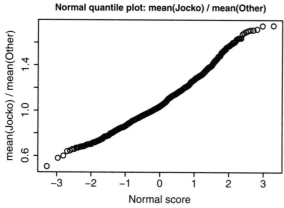

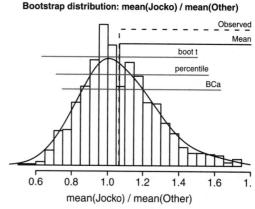

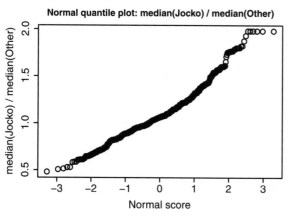

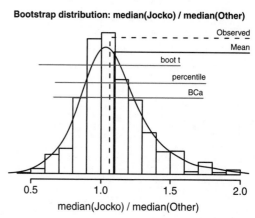

Chapter 17 Solutions

17.5. The center line is at $\mu = 85$ seconds. The control limits should be at $\mu \pm 3\sigma/\sqrt{6} = 85 \pm 3(17/\sqrt{6})$, which means about 64.18 and 105.82 seconds.

17.9. The most common problems are related to the application of the color coat; that should be the focus of our initial efforts.

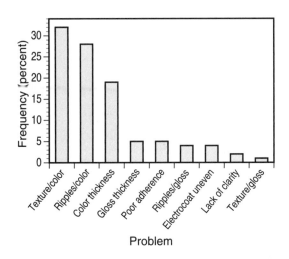

17.11. Possible causes could include delivery delays due to traffic or a train, high demand during special events, and so forth.

17.13. (a) For the $\bar{x}$ chart, the center line is at $\mu = 1.028$ lb; the control limits should be at $\mu \pm 3\sigma/\sqrt{3}$, which means about 0.9864 and 1.0696 lb. (b) For $n = 3$, $c_4 = 0.8862$ and $B_6 = 2.276$, so the center line for the s chart is $(0.8862)(0.024) = 0.02127$ lb, and the control limits are 0 and 0.05462 lb. (c) The control charts are below. (d) Both charts suggest that the process is in control.

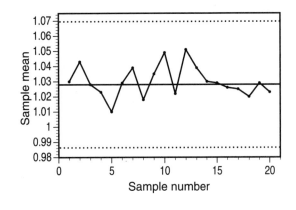 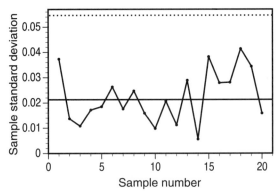

17.15. **(a)** The center line is at $\mu = 11.5$ Kp; the control limits should be at $\mu \pm 3\sigma/\sqrt{4} = 11.5 \pm 0.3 = 11.2$ and 11.8 Kp. **(b)** Graphs on the right and below. Points outside control limits are marked with an "X." **(c)** Set B is from the in-control process. The process mean shifted suddenly for Set A; it appears to have changed on about the 11th or 12th sample. The mean drifted gradually for the process in Set C.

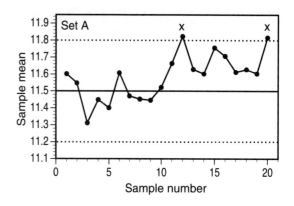

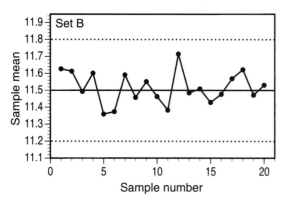

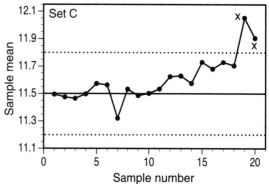

17.17. For the s chart with $n = 6$, we have $c_4 = 0.9515$, $B_5 = 0.029$ and $B_6 = 1.874$, so the center line is $(0.9515)(0.001) = 0.0009515$ inch, and the control limits are 0.000029 and 0.001874 inch. For the $\bar{x}$ chart, the center line is $\mu = 0.87$ inch, and the control limits are $\mu \pm 3\sigma/\sqrt{6} \doteq 0.87 \pm 0.00122 \doteq 0.8688$ and 0.8712 inch.

17.19. For the $\bar{x}$ chart, the center line is 43, and the control limits are 25.91 and 60.09. For $n = 5$, $c_4 = 0.9400$ and $B_6 = 1.964$, so the center line for the s chart is $(0.9400)(12.74) = 11.9756$, and the control limits are 0 and 25.02. The control charts (below) show that sample 5 was above the UCL on the s chart, but it appears to have been special cause variation, as there is no indication that the samples that followed it were out of control.

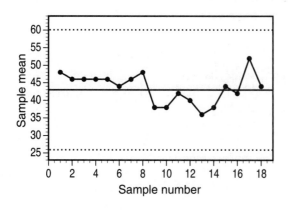

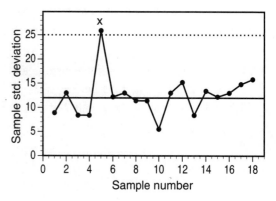

17.21. (a) The process mean is the same as the center line: $\mu = 715$. The control limits are three standard errors from the mean, so $30 = 3\sigma/\sqrt{4}$, meaning that $\sigma = 20$.
(b) If the mean changes to $\mu = 700$, then $\bar{x}$ is approximately Normal with mean 700 and standard deviation $\sigma/\sqrt{4} = 10$, so $\bar{x}$ will fall outside the control limits with probability $1 - P(685 < \bar{x} < 745) = 1 - P(-1.5 < Z < 4.5) = 0.0668$.
(c) With $\mu = 700$ and $\sigma = 30$, $\bar{x}$ is approximately Normal with mean 700 and standard deviation $\sigma/\sqrt{4} = 15$, so $\bar{x}$ will fall outside the control limits with probability $1 - P(685 < \bar{x} < 745) = 1 - P(-1 < Z < 3) = 0.16$.

17.23. The usual 3σ limits are $\mu \pm 3\sigma/\sqrt{n}$ for an $\bar{x}$ chart and $(c_4 \pm 3c_5)\sigma$ for an s chart. For 2σ limits, simply replace "3" with "2." **(a)** $\mu \pm 2\sigma/\sqrt{n}$. **(b)** $(c_4 \pm 2c_5)\sigma$.

17.25. (a) Shrinking the control limits would increase the frequency of false alarms, because the probability of an out-of-control point when the process is in control will be higher (roughly 5% instead of 0.3%). **(b)** Quicker response comes at the cost of more false alarms. **(c)** The runs rule is better at detecting gradual changes. (The one-point-out rule is generally better for sudden, large changes.)

17.27. We estimate $\hat{\sigma}$ to be $\bar{s}/0.9213 \doteq 1.1180$, so the $\bar{x}$ chart has center line $\bar{\bar{x}} = 47.2$ and control limits $\bar{\bar{x}} \pm 3\hat{\sigma}/\sqrt{4} \doteq 45.523$ and 48.877. The s chart has center line $\bar{s} = 1.03$ and control limits 0 and $2.088\hat{\sigma} \doteq 2.3344$.

17.29. One possible $\bar{x}$ chart is shown, created with the (arbitrary) assumption that the experienced clerk processes invoices in an average of 2 minutes, while the new hire takes an average of 4 minutes. (The control limits were set arbitrarily as well.)

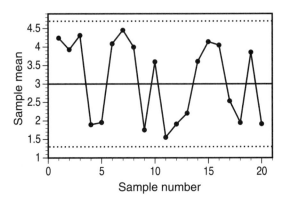

   Note: *Such a process would not be considered to be in control for very long. The initial control limits might be developed based on a historical estimate of σ, but eventually we should assess that estimate based on our sample standard deviations. Because both clerks "are quite consistent, so that their times vary little from invoice to invoice," each sample has a small value of s, so the revised estimate of σ would likely be smaller. At that point, the control limits (based on that smaller spread) will be moved closer to the center line.*

17.31. (a) Average the 20 sample means and standard deviations and estimate μ to be $\hat{\mu} = \bar{\bar{x}} = 2750.7$ and σ to be $\hat{\sigma} = \bar{s}/c_4 = 345.5/0.9213 \doteq 375.0$. **(b)** In the s chart shown in Figure 17.7, most of the points fall below the center line.

17.33. If the manufacturer practices SPC, that provides some assurance that the phones are roughly uniform in quality—as the text says, "We know what to expect in the finished product." So, assuming that uniform quality is sufficiently high, the purchaser does not need to inspect the phones as they arrive because SPC has already achieved the goal of that inspection: to avoid buying many faulty phones. (Of course, a few unacceptable phones may be produced and sold even when SPC is practiced—but inspection would not catch all such phones anyway.)

17.35. The quantile plot does not suggest any serious deviations from Normality, so the natural tolerances should be reasonably trustworthy.

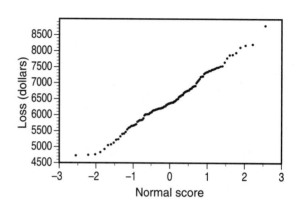

Note: *We might also assess Normality with a histogram or stemplot; this looks reasonably Normal, but we see that the number of losses between $6000 and $6500 is noticeably higher than we might expect from a Normal distribution. In fact, the smallest and largest losses were $4727 and $8794. These are both within the tolerances, but note that the minimum is quite a bit more than the lower limit of the tolerances ($4008). The large number of losses between $6000 and $6500 makes the mean slightly lower and therefore lowers both of the tolerance limits.*

17.37. If we shift the process mean to 2500 mm, about 99% will meet the new specifications:
$$P(1500 < X < 3500) = P\left(\tfrac{1500-2500}{383.8} < Z < \tfrac{3500-2500}{383.8}\right) = P(-2.61 < Z < 2.61) \doteq 0.9910$$

17.39. The mean of the 17 in-control samples is $\bar{\bar{x}} = 43.4118$, and the standard deviation is 11.5833, so the natural tolerances are $\bar{\bar{x}} \pm 3s = 8.66$ to 78.16.

17.41. Only about 44% of meters meet the specifications. Using the mean (43.4118) and standard deviation (11.5833) found in the solution to Exercise 17.39:
$$P(44 < X < 64) = P\left(\tfrac{44-43.4118}{11.5833} < Z < \tfrac{64-43.4118}{11.5833}\right) = P(0.05 < Z < 1.78) \doteq 0.4426$$

17.43. The limited precision of the measurements shows up in the granularity (stair-step appearance) of the graph. Aside from this, there is no particular departure from Normality.

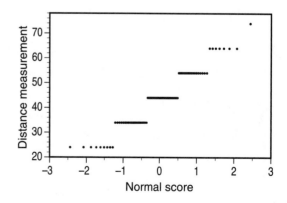

17.45. The quantile plot, while not perfectly
linear, does not suggest any serious deviations from Normality, so the natural
tolerances should be reasonably trustworthy.

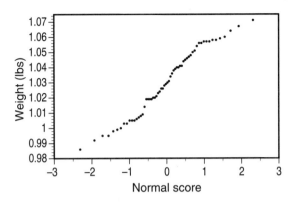

17.47. (a) (ii) A sudden change in the $\bar{x}$ chart: This would immediately increase the amount of
time required to complete the checks. **(b)** (i) A sudden change (decrease) in s or R because
the new measurement system will remove (or decrease) the variability introduced by human
error. **(c)** (iii) A gradual drift in the $\bar{x}$ chart (presumably a drift up, if the variable being
tracked is the length of time to complete a set of invoices).

17.49. The process is no longer the same as it was during the downward trend (from the
1950s into the 1980s). In particular, including those years in the data used to establish the
control limits results in a mean that is too high to use for current winning times, and a
standard deviation that includes variation attributable to the "special cause" of the changing
conditioning and professional status of the best runners. Such special cause variation should
not be included in a control chart.

17.51. LSL and USL are specification limits on the individual observations. This means that
they do not apply to averages and that they are *specified* as desired output levels, rather than
being *computed* based on observation of the process. LCL and UCL are control limits for
the averages of samples drawn from the process. They may be determined from past data,
or independently specified, but the main distinction is that the purpose of control limits is to
detect whether the process is functioning "as usual," while specification limits are used to
determine what percentage of the outputs meet certain specifications (are acceptable for use).

17.53. For computing $\hat{C}_{pk}$, note that the estimated process mean (2750.7 mm) lies closer
to the USL. **(a)** $\hat{C}_p = \dfrac{4000 - 1000}{6 \times 383.8} \doteq 1.3028$ and $\hat{C}_{pk} = \dfrac{4000 - 2750.7}{3 \times 383.8} \doteq 1.0850$.
(b) $\hat{C}_p = \dfrac{3500 - 1500}{6 \times 383.8} \doteq 0.8685$ and $\hat{C}_{pk} = \dfrac{3500 - 2750.7}{3 \times 383.8} \doteq 0.6508$.

17.55. In the solution to Exercise 17.44, we found that the mean and standard deviation of all
60 weights are $\bar{x} \doteq 1.02996$ lb and $s \doteq 0.0224$ lb. **(a)** $\hat{C}_p = \dfrac{1.10 - 0.94}{6 \times 0.0224} \doteq 1.1901$ and
$\hat{C}_{pk} = \dfrac{1.10 - 1.03}{3 \times 0.0224} \doteq 1.0418$. (These were computed with the unrounded values of $\bar{x}$ and s;
rounding will produce slightly different results.) **(b)** Customers typically will not complain
about a package that was too heavy.

17.57. (a) $C_{pk} = \frac{0.75 - 0.25}{3\sigma} \doteq 0.5767$. 50% of the output meets the specifications. **(b)** LSL and USL are 0.865 standard deviations above and below to mean, so the proportion meeting specifications is $P(-0.865 < Z < 0.865) \doteq 0.6130$. **(c)** The relationship between C_{pk} and the proportion of the output meeting specifications depends on the shape of the distribution.

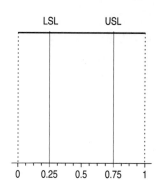

17.59. See also the solution to Exercise 17.43. **(a)** Use the mean and standard deviation of the 85 remaining observations: $\hat{\mu} = \bar{x} = 43.4118$ and $\hat{\sigma} = s = 11.5833$. **(b)** $\hat{C}_p = \frac{20}{6\hat{\sigma}} \doteq 0.2878$ and $\hat{C}_{pk} = 0$ (because $\hat{\mu}$ is outside the specification limits). This process has very poor capability: The mean is too low and the spread too great. Only about 46% of the process output meets specifications.

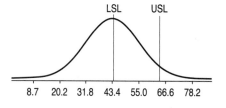

17.61. We have $\bar{x} = 22.005$ mm and $s = 0.009$ mm, so we assume that an individual bearing diameter X follows a $N(22.005, 0.009)$ distribution. **(a)** About 85.3% meet specifications:

$$P(21.985 < X < 22.015) = P\left(\frac{21.985 - 22.005}{0.009} < Z < \frac{22.015 - 22.005}{0.009}\right)$$
$$= P(-2.22 < Z < 1.11)$$
$$= 0.9868 - 0.1335 = 0.8533.$$

(b) $\hat{C}_{pk} = \frac{22.015 - 22.005}{3 \times 0.009} \doteq 0.3704$.

17.63. This graph shows a process with Normal output and $C_p = 2$. The tick marks are σ units apart; this is called "six-sigma quality" because the specification limits are (at least) six standard deviations above and below the mean.

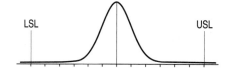

17.65. Students will have varying justifications for the sampling choice. Choosing six calls per shift gives an idea of the variability and mean for the shift as a whole. If we took six consecutive calls (at a randomly chosen time), we might see additional variability in $\bar{x}$ because sometimes those six calls might be observed at particularly busy times (when a customer has to wait for a long time until a representative is available or when a representative is using the restroom).

17.67. The outliers are 276 seconds (sample 28), 244 seconds (sample 42), and 333 seconds (sample 46). After dropping those outliers, the standard deviations drop to 9.284, 6.708, and 31.011 seconds. (Sample #39, the other out-of-control point, has two moderately large times, 144 and 109 seconds; if they are removed, s drops to 3.416.)

17.69. (a) For those 10 months, there were 1028 overdue invoices out of 28,400 total invoices (opportunities), so $\bar{p} = \frac{1028}{28,400} \doteq 0.03620$. **(b)** The center line and control limits are:

$$CL = \bar{p} = 0.03620, \text{control limits: } \bar{p} \pm 3\sqrt{\frac{\bar{p}(1-\bar{p})}{2840}} = 0.02568 \text{ and } 0.04671$$

17.71. The center line is at the historical rate (0.0189); the control limits are

$0.0189 \pm 3\sqrt{\frac{0.0189 \cdot 0.9811}{500}}$, which means about 0.00063 and 0.03717.

17.73. The center line is at $\bar{p} = \frac{163}{36,480} \doteq 0.004468$; the control limits should be at

$\bar{p} \pm 3\sqrt{\frac{\bar{p}(1-\bar{p})}{1520}}$, which means about -0.00066 (use 0) and 0.0096.

17.75. (a) The student counts sum to 9218, while the absentee total is 3277, so $\bar{p} = \frac{3277}{9218} = 0.3555$ and $\bar{n} = 921.8$. **(b)** The center line is $\bar{p} = 0.3555$, and the control limits are:

$$\bar{p} \pm 3\sqrt{\frac{\bar{p}(1-\bar{p})}{921.8}} = 0.3082 \text{ and } 0.4028$$

The p chart suggests that absentee rates are in control. **(c)** For October, the limits are 0.3088 and 0.4022; for June, they are 0.3072 and 0.4038. These limits appear

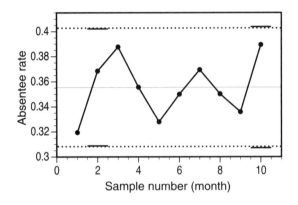

as solid lines on the p chart, but they are not substantially different from the control limits found in (b). Unless n varies *a lot* from sample to sample, it is sufficient to use $\bar{n}$.

17.77. (a) $\bar{p} = \frac{8000}{1,000,000} = 0.008$. We expect about $4 = (500)(0.008)$ defective orders per month. **(b)** The center line and control limits are:

$$CL = \bar{p} = 0.008, \text{control limits: } \bar{p} \pm 3\sqrt{\frac{\bar{p}(1-\bar{p})}{500}} = -0.00395 \text{ and } 0.01995$$

(We take the lower control limit to be 0.) It takes at least ten bad orders in a month to be out of control because $(500)(0.01995) = 9.975$.

17.79. **(a)** The percents do not add to 100% because one customer might have several complaints; that is, he or she could be counted in several categories. **(b)** Clearly, top priority should be given to the process of creating, correcting, and adjusting invoices, as the three most common complaints involved invoices.

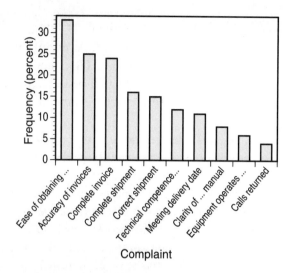

17.81. On one level, these two events are similar: Points below the LCL on an $\bar{x}$ (s) chart suggest that the process mean (standard deviation) may have decreased. The difference is in the implications of such a decrease (if not due to a special cause). For the mean, a decrease might signal a need to recalibrate the process in order to keep meeting specifications (that is, to bring the process back into control). A decrease in the standard deviation, on the other hand, typically does not indicate that adjustment or recalibration is necessary, but it will require re-computation of the $\bar{x}$ chart control limits.

17.83. We find that $\bar{s} = 7.65$, so with $c_4 = 0.8862$ and $B_6 = 2.276$, we compute $\hat{\sigma} = 8.63$ and UCL = 19.65. One point (from sample #1) is out of control. (And, if that cause were determined and the point removed, a new chart would have s for sample #10 out of control.) The second (lower) UCL line on the control chart is the final UCL, after removing both of those samples (per the instructions in Exercise 17.84).

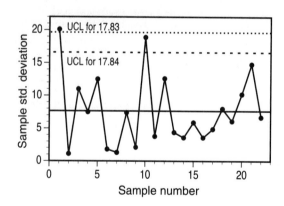

17.85. **(a)** As was found in the previous exercise, $\hat{\sigma} = \bar{s}/c_4 \doteq 7.295$. Therefore, $C_p = \frac{50}{6\hat{\sigma}} \doteq 1.1423$. This is a fairly small value of C_p; the specification limits are just barely wider than the $6\hat{\sigma}$ width of the process distribution, so if the mean wanders too far from 830, the capability will drop. **(b)** If we adjust the mean to be close to 830 mm $\times$ 10^{-4} (the center of the specification limits), we will maximize C_{pk}. C_{pk} is more useful when the mean is not in the center of the specification limits. **(c)** The value of $\hat{\sigma}$ used for determining C_p was estimated from the values of s from our control samples. These are for estimating short-term variation (within those samples) rather than the overall process variation. To get a better estimate of the latter, we should instead compute the standard deviation s of the *individual* measurements used to obtain the means and standard deviations given in Table 17.11 (specifically, the 60 measurements remaining after dropping samples 1 and 10). These numbers are not available. (See "How to cheat on C_{pk}" on page 17–44 of Chapter 17.)

17.87. (a) Use a p chart, with center line $\bar{p} = \frac{15}{5000} = 0.003$ and control limits $\bar{p} \pm 3\sqrt{\dfrac{\bar{p}(1-\bar{p})}{100}}$, or 0 to 0.0194. (b) There is little useful information to be gained from keeping a p chart: If the proportion remains at 0.003, about 74% of samples will yield a proportion of 0, and about 22% of proportions will be 0.01. To call the process out of control, we would need to see two or more unsatisfactory films in a sample of 100.

17.89. Several interpretations of this problem are possible, but for most reasonable interpretations, the probability is about 0.3%. From the description, it seems reasonable to assume that all three points are inside the control limits; otherwise, the one-point-out rule would take effect. Furthermore, the phrase "two out of three" could be taken to mean either "*exactly* two out of three," or "*at least* two out of three." (Given what we are trying to detect, the latter makes more sense, but students may have other ideas.)

For the kth point, we name the following events:
- A_k = "that point is no more than $2\sigma/\sqrt{n}$ from the center line,"
- B_k = "that point is 2 to 3 standard errors from the center line."

For an in-control process, $P(A_k) = 95\%$ (or 95.45%) and $P(B_k) = 4.7\%$ (or 4.28%). The first given probability is based on the 68–95–99.7 rule; the second probability (in parentheses) comes from Table A or software.

Note that, for example, the probability that the first point gives no cause for concern, but the second and third are more than $2\sigma/\sqrt{n}$ from, and on the same side of, the center line, would be:

$$\tfrac{1}{2}P(A_1 \cap B_2 \cap B_3) \doteq 0.10\% \ (\text{or } 0.09\%)$$

(The factor of 1/2 accounts for the second and third points being on the same side of the center line.) If the "other" point is the second or third point, this probability is the same, so if we interpret "two out of three" as meaning "*exactly* two out of three," then the total probability is three times the above number:

$$P(\text{false out-of-control signal from an in-control process}) \doteq 0.31\% \ (\text{or } 0.26\%)$$

With the (more-reasonable) interpretation "*at least* two out of three":

$$P(\text{false out-of-control signal}) = \tfrac{1}{2}P(A_1 \cap B_2 \cap B_3) + \tfrac{1}{2}P(B_1 \cap A_2 \cap B_3) + \tfrac{1}{2}P(B_1 \cap B_2)$$
$$\doteq 0.32\% \ (\text{or } 0.27\%)$$